YOUR BIRTHDAY
IN HISTORY

YOUR BIRTHDAY
IN HISTORY

By Beatrice Nikola, B.A., B.S., J.D.

Founded 1910
THE CHRISTOPHER PUBLISHING HOUSE
HANOVER, MASSACHUSETTS 02339

COPYRIGHT © 2000
BY BEATRICE NIKOLA, B.A., B.S., J.D.

Library of Congress Catalog Number 98-73835

ISBN: 0-8158-0537-3

Publisher is not responsible for the
accuracy of data compiled in this publication.

PRINTED IN THE UNITED STATES OF AMERICA

Dedicated to my granddaughters,
Janelle Marie and Anne

January 1

- 1502 Pope Gregory XIII (Ugo Boncompani). Author of Gregorian Calendar
- 1526 Louis Bertrand
- 1655 Christian Thomasius. Philosopher; writer
- 1705 Charles Councy
- 1726 Jacques-Denis Cochin. Writer
- 1735 Paul Revere. Participated in Boston Tea Party; rode to Concord to warn that the British were coming
- 1745 General Anthony Wayne. Served during Revolutionary War at Brandywine, Germantown and Stony Point
- 1752 Betsy Ross. Made the first U.S. flag
- 1767 Maria Edgeworth. Novelist
- 1781 Capt. Robert Lucas. Surveyer; senator; governor of Ohio
- 1802 Martin Van Hook
- 1820 Lester Wallach. Actor
- 1834 Ludovic Halevy. Novelist
- 1839 James Ryder Randall. Poet
- 1840 Patrick Walsh. Journalist; writer; editor; senator
- 1857 Maggie Cline
- 1858 Albert Gleaves. Admiral; governor of Naval Home at Philadelphia; writer
- 1878 Edwin Goldman. Conductor; founder, American Bandmaster's Association
- 1879 Edward Forster. Writer
- 1883 "Wild Bill" Donovan
- 1895 J. Edgar Hoover. Director, F.B.I.
- 1898 Marta Fuchs. Soprano
- 1908 Bob Nolan of "The Sons of the Pioneers"
- 1909 Senator Barry Goldwater
- 1910 Elsie Sebelius of Lichtenstein and the U.S.
- 1911 Hank Greenberg. All-time baseball great
- 1912 Dana Andrews. Actor
- 1914 Lloyd Mangrum. Pro golfer
- 1923 Milt Jackson
- 1924 Earl Torgeson of Boston Braves, Philadelphia Phillies, Detroit Tigers, Chicago White Sox and New York Yankees
- 1925 Valentina Cortesa. Actress
- 1936 Eve Queler. Conductor
- 1938 Picko Troberg. Race car driver; winner of Grand Touring Title

1945 Tom Selleck
1950 Steve Barkett. Actor

January 2

1403 Bessarion. Fifteenth Century Greek Bishop
1512 Christobal Morales. Composer
1647 Nathaniel Bacon
1657 Stephen Bryant
1658 Hannah Kinney
1729 Jacob Benton
1752 John Laugee
1752 Philip Freneau. Poet
1758 Joseph Cross of Revolutionary War fame
1762 James Corbin. Patriot
1790 Tempe Skinner
1809 Friedrich Johns. Composer
1822 Rudolf Clausius. Founder of thermodynamics
1830 Henry Kingsley. Writer; editor
1832 William O'Connor. Writer
1840 John Lancaster Spaulding
1853 George Angier Gordon
1867 August Benziger. Painter
1870 Ernst Barlach. Sculptor
1873 St. Theresa of Lisieux
1876 William Jeffers. Railroader
1881 Frederick Varley. Painter
1886 Florence Laurence. Actress
1887 August Mache. Painter
1894 Robert Nathan. Novelist
1904 Sally Rand
1910 Ulrich Becker. Novelist
1916 Robert Newmann. Ambassador
1917 Vera Zorina. Actress
1934 Edwin Apfel. Writer
1936 Roger Miller. Singer; guitarist
1938 Robert Smithson. Artist
1963 Tammy Warn

January 3

106BC Marcus Tullius Cicero. Orator
1504 Gian Albani. Statesman

1618	Jean Crasset. Writer
1694	St. Paul of the Cross. Missionary; martyr
1698	Pietro Metastasio. Sculptor
1719	Friedrich Ethal
1729	Edward Winslow
1759	Theophilus Lovering of Revolutionary War
1769	Moses Chenay
1776	Thomas Morris. Senator
1800	Etienne Faillon
1823	Robert Whitehead. Inventor
1840	Ven. Joseph de Veuster (Father Damien). Missionary to leper colony of Molokai
1884	Mary Hamilton Swindler
1886	John Fletcher. Writer
1886	Josephine Hull. Actress
1892	J.R.R. Tolkien. Author
1900	Robert Adair. Actor
1900	Zasu Pitts
1905	Ray Milland
1909	Victor Borge
1919	Huschke von Hanstein. Race car driver
1921	John Russell. Actor
1929	Bobby Hull of pro hockey

January 4

1684	Jakob Balde
1642	Philippe Pierson
1643	Sir Isaac Newton
1676	Louis Le Moyne. Hero of the capture of Fort Monsipi
1679	Roger Walcott. Colonial governor of Connecticut
1682	Jacopo Facciolati. Lexicographer
1710	Giovanni Battista Pergolesi. Composer
1727	Hezekiah Blanchard
1746	Benjamin Rush. Signer of Declaration of Independence
1764	Richard Christmas of Revolutionary War
1785	Jacob Grimm of the Grimm Brothers, authors of Grimm Fairy Tales
1809	Louis Braille. Inventor of Braille System
1818	Pat Wittman. Journalist
1858	Carter Glass. Congressman; senator; Secretary of Treasury

1870	Percy Pitt. Conductor
1878	Augustus John. Artist
1881	Wilhelm Lehmbruck. Sculptor
1889	Maisie Ward. Co-founder of publishing house; author
1895	Leo Grumman. Aircraft pioneer
1899	James Haley. Statesman
1914	Jane Wyman
1935	Floyd Patterson
1937	Grace Bumbry. Singer
1937	Dyan Cannon
1941	Kermit Alexander of the Rams

January 5

1548	Francisco Suarez. Scholar
1592	Shah Johan. Builder of the Taj Mahal
1736	Pierre Constantin
1744	Gaspar Jovellanos. Statesman
1746	Ezekiel Bailey
1767	Jean Baptiste Say. Economist
1778	Zebulon Pike for whom Pike's Peak is named
1779	Stephen Decatur, Jr. Naval hero; commander of the "Enterprise"
1791	Benjamin Gildersleeve
1824	Utto Kornmuller. Composer
1833	Eugene Hilgard
1862	Christopher La Farge
1876	Konrad Adenauer
1884	Edward Frankland. Novelist
1900	Yves Tangry. Painter
1906	Kathleen Kenyon. Archeologist
1911	Jeanne-Pierre Aumont
1926	William De Witt Snodgrass. Poet
1928	Raylin Moore. Writer
1931	Alvin Ailey
1931	Alfred Brendel. Pianist
1931	Robert Duval. Actor
1938	King Juan Carlos of Spain
1946	Diane Keaton
1947	Kathy Switzer. Marathon runner
1960	Tim Kerr of hockey

January 6

- 1412 St. Joan of Arc
- 1486 Martin Agricola. Composer
- 1500 Bl. John of Avila
- 1563 Martin Becanus. Writer
- 1568 Henri Spondamus
- 1633 Isaac Stearns
- 1735 James Dabney
- 1749 Vittorio Alfieri. Poet
- 1765 Bartholomew Clute
- 1794 Mother Frances Mary Teresa Ball. Foundress
- 1799 Jadedish Strong Smith. Explorer
- 1811 Senator Charles Sumner for whom Fort Sumner was named
- 1842 Walter Elliott. Missionary; writer
- 1858 Benjamin Davis. Tenor
- 1859 Adm. Hugh Rodman
- 1859 Senator Duncan Upshaw Fletcher. Co-sponsored the Act which created the Securities and Exchange Commission
- 1880 Tom Mix
- 1882 Sam Rayburn. Speaker of the House longer than anyone in history
- 1886 Russell Waesche. Commander of U.S. Coast Guard in World War II
- 1891 Arthur Albertson. Actor
- 1900 Kathryn Hulme. Author
- 1903 Boris Blacher. Composer
- 1910 Wright Morris. Writer
- 1913 Tom Brown. Actor
- 1913 Vince Lombardi
- 1913 Loretta Young
- 1914 Danny Thomas
- 1921 Louis Harris of Harris Poll
- 1931 Edgar Doctorow. Novelist

January 7

- 1729 Jonathan Cowherd of the Revolutionary War
- 1753 Benjamin Racer
- 1794 Eilhard Mitcherlich. Scientist
- 1800 Millard Fillmore. President of the United States

1809	Jesse Winter
1813	Ann Rutledge
1824	Eliza Jan Farnham
1824	Julia Kavanagh
1830	Joseph Kirkland. Writer
1832	Thomas De Witt Talmadge
1833	St. Bernadette of Lourdes
1847	Casper Goodrich. Admiral; founder of the Naval War College and President of Naval History Society
1861	Louise Guiney. Poet
1862	Mary Wilkins. Novelist
1865	Lyman Duff. Chairman of the Duff Commission
1873	Adm. Calvin Durgun. Commander of the "Ranger"; Deputy Chief of Naval Operations
1899	Joseph Lee. Governor of Utah
1899	Francis Paulene. Composer
1901	John Brownlee. Baritone
1901	Zora Hurston. Writer
1910	Orval Faubus. Governor of Arkansas
1912	Charles Addams. Cartoonist
1919	Robert Duncan. Poet
1924	Anne Vernon. Actress
1928	Clyde Snow. Anthropologist
1934	Ralph Allen. Producer
1934	Joe Ponazecki. Actor
1941	Harvey Evans. Actor

January 8

1589	Ivan Gundulic. Poet
1735	John Carroll. First archbishop of Baltimore
1753	Benjamin Cushman
1809	Nikolaus Becker. Poet
1821	James Longstreet. Confederate general at Bull Run
1823	Florent Willems. Painter
1824	Wilkie Collins. Novelist
1825	William Brady
1830	Hans von Bulow. Pianist
1843	Lucille Western. Actress
1857	Augustus Thomas. Actor

1862 Joseph Dechelette. Author
1884 Adm. John J. Brady. Recipient of Navy Cross and Distinguished Service Cross
1894 St. Maximilian Kolbe. Martyr of concentration camp
1898 Revven Kosakoff. Violinist
1902 Hugh Hencken. Archeologist
1904 Peter Arnold. Cartoonist
1912 José Ferrer
1917 Peter Taylor. Writer
1925 Polly Page. Author
1934 Jacques Anquetil. Cyclist
1934 Maud Gatewood. Artist
1947 David Bowie. Actor
1958 Amy Ryder
1963 Tina Belis of Broadway
1965 Jacquelyn Piro

January 9

1658 Nicolas Coustou. Sculptor
1674 Reinhard Keiser. Composer
1699 Robert Pothier
1724 Isaac Backus
1726 Nehemiah Lovewell of Revolutionary War
1798 Israel Lawton
1803 Christopher Memminger. Secretary of the Treasury of the Confederate States of America
1805 Charles Gayarré. Historian
1810 Anna Bishop. Singer
1819 William Frith. Painter
1837 Julius Burroughs. Senator
1839 John Knowles Paine
1848 Admiral William Kimball. Commander of Nicaraguan Expedition
1849 Markus Backman. Knighted by Haakon VII of Denmark
1854 Jennie Churchill. Musician; essayist; editor
1858 Elizabeth Brittain. Botanist
1873 Chaim Bialik. Poet
1875 Gertrude Whitney. Sculptor
1879 Emory Land. Maritime chairman; writer
1881 Lascelles Abercrombie. Writer

1894	Edward Moore. Author
1902	Monsignor Jose Maria Escriva Balguer. Author; founder of Opus Dei
1913	Richard Nixon. President of the United States
1914	Gypsy Rose Lee
1919	William Meredith. Poet
1928	Judith Krantz
1951	Crystal Gayle

January 10

1480	Margaret of Austria. Regent of the Netherlands
1607	St. Isaac Joques. Missionary; North American Martyr
1654	Joshua Barnes. Writer
1710	Sarah Butler
1729	Lazzaro Spallanzani. Scientist
1734	Thomas Crandall
1738	Ethan Allen. Captured Fort Ticonderoga
1757	Edward Crawford of the Revolutionary War
1764	William Christy. Major of the Louisiana Rangers; registrar of the United States Land Office
1795	Francis Brunner. Missionary
1805	Napoleon Joseph Perche
1814	Aubrey de Vere. Poet
1820	Louisa Drew. Actress
1825	Elbridge Stevens
1834	John Acton. Historian
1840	Louis Nazaire Begin
1880	John Root. Architect
1884	James Philip Dunn. Composer
1886	José Donostia. Composer
1887	Robinson Jeffers. Poet
1889	John Hold. Cartoonist
1903	Barbara Hepworth
1904	Ray Bolger. Actor
1921	Roger Ward. Race car driver
1930	Sal Mineo. Actor

January 11

347	Theodosius I. Roman Emperor
1503	Francesco Mazzola. Painter

1737	Richard Salter. Patriot
1757	Alexander Hamilton
1762	Northend Cogswell. Patriot
1801	Caralina Kirkland. Author
1814	Sir James Paget. Founder of pathology
1815	John A. Macdonald. First prime minister of Canada
1819	Francis Seelas. Missionary
1827	George Cannon. Polygamist who was expelled from Congress
1838	Ivan Cagliero
1841	Otto Gierke. Historian
1859	Homer Parsons
1858	Cave Johnson. Postmaster General; introduced postage stamps
1870	Alice Rice. Author
1876	Benedict Brown
1885	Admiral Worrall Carter. Writer
1886	Aldo Leopold. Naturalist
1892	Francis Xavier Ford. Martyr
1905	Frederic Cannay. Co-author of *Ellery Queen* series
1905	Joseph Doherty. First bishop of Yakima
1912	Hugh Stubbins. Architect
1945	Ben Nielson. Mayor of Fountain Valley
1963	Tracy Caulkins. Gold Medal winner

January 12

1588	John Winthrop
1644	Nathaniel Cheney
1662	Samuel Shute. Colonial governor of Massachusetts
1716	Antonis de Ulla. Scientist
1729	Edmund Burke
1753	George Baylor of Revolutionary War
1763	George Michel. Painter
1771	Jacob Lenz. Poet
1801	Paul Taglioni. Choreographer
1837	Thomas Moran. Painter
1841	James Harahn. President of Illinois Central
1843	Gen. George Burton of Modoc Indian War
1856	John Sargent. Painter
1859	Henry Heitfeld. Senator; mayor of Lewiston
1864	Annie Russell. Actress

1876	Jack London
1878	Ference Molnar. Author
1879	Ray Harraun. Race driver
1882	Milton Sills. Actor
1888	John Bernard Kelly. Author; organized Golden Book Awards
1899	Fritz Crisler of National Football Hall of Fame
1907	Tex Ritter
1908	José Limon. Dancer
1908	Leopold Ludwig. Conductor
1912	Pat Hurley. Scientist
1928	Lloyd Ruby. Racing driver
1938	Alan Rees of racing fame
1950	Randy Jones of baseball

January 13

1381	St. Colette
1628	Charles Perrault. Poet
1683	Christoph Graupner. Composer
1705	Elizabeth Blanchard of Woburn
1722	Charles Walmesley
1753	Augustus Belding
1763	Matilda Gates of Marlboro
1768	Samuel Whipple
1772	John Lathrop. Early settler
1808	Salmon Portland Chase. Chief Justice
1813	Nathaniel Bannister. Actor
1819	Malumba Ary
1827	Ethel Beers. Poet
1834	Horatio Alger
1847	John Atkinson Moore. Merchant of Vermont
1884	Sophie Tucker
1885	Alfred Fuller. The original Fuller Brush Man
1889	Earl Zeigler. Recipient of Blackstone Fellowship
1890	Elmer Davis. News Analyst
1903	Charles Kullman. Tenor
1919	Robert Stack
1941	Faye Dunaway
1948	George Adams

January 14

1624	Henri Boudon. Writer
1683	Gottfried Silbermann. Piano builder
1730	William Whipple. Signer of Declaration of Independence
1749	James Garrard. Governor of Kentucky
1751	Joseph Lum of the Revolution
1799	John Turner. Railroad pioneer
1780	Henry Baldwin. Supreme Court Justice
1802	Leon Halevy. Poet
1806	Matthew Maury. Oceanographer
1809	Boniface Wimmer. Missionary
1816	Francis Kernan. Congressman; senator
1820	Mary Airhart
1831	William Washburn. U.S. Surveyor General
1835	Frederic Colin. Missionary
1841	Berthe Morisat. Painter
1847	Bordon Parker Bourne. Editor
1848	James Sediner
1854	Benjamin Odell. Governor
1863	Richard Outcault. Artist
1875	Albert Schweitzer
1878	Isabel Maitland Stewart
1890	Arthur Holmer. Scientist
1899	Nevil Shute. Writer
1902	Alfred Tarski
1902	Moses Zucker. Author
1905	Emily Hahn. Writer
1906	William Bendix
1918	James Barnes. Musician
1928	Gerald Arpino. Choreographer
1932	Harriet Andersson. Actress
1941	Faye Dunaway

January 15

1432	Afonso V. King of Portugal
1716	Philip Livingston. Signer of Declaration of Independence
1744	Nathaniel Gubtail. Patriot
1753	Richard Whittington of the Revolutionary War
1759	Jean Baptiste Rebalair

1770	Ludwig von Beethoven
1785	William Prout. Scientist
1793	Ferdinand Waldmüller. Painter
1811	St. Joseph Cafasso
1812	Alexander Thomson
1821	John Breckenridge. Vice President of the United States
1835	Gen. Patrick Guiney of the Battle of Chickahominy of Civil War
1836	Constance Runcie. Composer
1841	Charles Briggs. Editor
1845	Heinrich Vogl. Tenor
1859	Nathaniel Britton. Botanist
1862	Lole Fuller. Actress
1893	Matthew Brady
1903	Andrew Dever. Governor of Massachusetts
1908	Edward Teller. Physicist
1913	Lloyd Bridges
1919	Bill Stroppe. Race car driver
1937	Margaret O'Brien
1944	Ronnie Milsap

January 16

1724	Samuel Fitch. Banished from Boston
1728	Nicollo Piccini. Composer
1737	Samuel Cowell. Patriot
1749	Vittorio Alfieri. Poet
1763	François Talma. Actor
1783	Jean Bouvier
1801	Cornelius Smock
1807	Charles Davis. Admiral and scientist
1809	Ebenezer Grubb
1815	Henry Halleck. Union general of Civil War
1845	Adm. Sigsbee of Battles of Mobile Bay and Fort Fisher
1853	André Michelin of tire fame
1858	William Pickering. Astronomer
1863	Paul Wattson. Established "Graymoor Press"
1864	Frank Bacon. Actor
1874	Robert Service. Writer
1881	Joseph Reilly
1904	Jo-Jo Morrissey of baseball

1907 Hubert Creekmore. Author
1909 Ethel Merman
1911 Dizzy Dean of baseball
1928 Eartha Kitt
1934 Marilyn Horne
1935 A.J. Foyt. Race car driver

January 17

1342 Philip the Bold. Duke of Burgundy
1463 Antoine Duprat
1504 Pope St. Pius V (Antony Ghislieri)
1560 Gaspard Bauhin. Scientist
1706 Benjamin Franklin
1734 François Gassec. Composer
1753 Abraham Bennet of Revolutionary War
1757 Freeborn Sweet. Patriot
1771 Charles Brockden Brown. Novelist
1794 Fanny Phelps
1837 François Lenormant
1848 Louis Tiffany. Designer of lamps and vases
1851 A.B. Frost. Cartoonist
1860 Douglas Hyde
1874 Jean Calvet
1876 Frank Hague. Mayor of Jersey City
1880 Mack Sennett. Comedian
1882 William Conrad Bruce. Founder of Serra International
1897 Nils Asther. Actor
1899 Al Capone
1907 Henk Badings. Composer
1920 Nora Kaye. Dancer
1926 Moira Shearer. Actress
1927 Dr. Thomas Dooley
1928 Jean Barraque. Composer
1929 Jaques Plante of the National Hockey League
1931 James Earl Jones. Actor
1937 Troy Donahue
1942 Cassius Marcellus Clay (Muhammad Ali). Heavyweight Champion
1963 Matthew Paul of bowling fame

January 18

- 1732 Lucy Calmes
- 1761 Newcomb Kinney
- 1782 Daniel Webster
- 1782 John Calhoun
- 1809 Commodore Oliver Glisson
- 1813 George Rex Graham
- 1818 Richard Yates. Governor of Illinois
- 1825 George W. Hutton
- 1826 Barnett Washington James
- 1835 César Cui. Composer
- 1841 Emmanuel Chalrier. Composer
- 1844 Walter Satterlie. Painter
- 1848 Matthew Webb. Pro swimmer
- 1850 Seth Low. Mayor of New York
- 1856 Dr. Daniel Hale Williams. Medical pioneer
- 1867 Ruben Dario. Poet
- 1882 Alan Milne. Writer
- 1892 Oliver Hardy. Actor
- 1899 Albert Murray
- 1904 Cary Grant
- 1904 Robert Thomas Daniel. Writer
- 1907 George Granville Haney. Pioneer
- 1913 Danny Kaye
- 1930 Evelyn Lear. Vocalist
- 1941 Bobby Goldsboro
- 1942 Johnny Servaz-Gairn. Race car driver

January 19

- 1576 Jean Arnoux
- 1668 Jean Vincennes. Explorer
- 1702 Abigail Willard. Descendant of Priscilla and John Alden
- 1736 James Watt. Developer of steam engine
- 1747 Johann Bade. Astronomer
- 1749 Isaiah Thomas. Founder of American Antiquarian Society
- 1782 Daniel Auber. Composer
- 1787 Mary Aikenhead. Foundress of the Irish Sisters of Charity
- 1789 Dr. Ruben Palmer. Surgeon of War of 1812

1807	William Newman. Senator
1807	General Robert E. Lee
1809	Edgar Allen Poe
1819	George Washington Lewis of Seminole and Mexican Wars
1820	Bennet Engle of Harper's Ferry
1820	Harriet Mathilda Helmerhausen. Pioneer
1839	Paul Cézanne. Artist
1847	Josiah Strong
1853	Stephen White. Senator
1889	Theodore Roener. Historian
1905	Oveta Culp Hobby
1906	Philo Taylor Farnsworth. Television pioneer
1906	Lanny Ross. Singer
1909	Adm. Lawson Ramage. Recipient of Medal of Honor
1921	Patricia Highsmith. Novelist
1920	Jean Stapleton
1922	Guy Madison
1931	Ron Packard
1946	Dolly Parton

January 20

1502	Bl. Sebastian De Aparicio
1554	Sebastian. King of Portugal
1573	Simon Mayr. Scientist
1716	Carlos III. King of Spain
1736	Abner Crowell
1752	Lemuel Crooker. Patriot
1760	Ephraim Blackford of the Revolutionary War
1782	John of Austria
1798	Anson Jones. Last president of the Republic of Texas
1806	Nathaniel Willis. Author
1814	David Wilmat. Congressman; senator
1820	Ebenezer Jones. Poet
1829	Thomas Bridgett. Author
1832	Frances Margaret Taylor. Foundress of the Poor Servants of the Mother of God
1832	William Larrabee. Governor of Iowa
1848	Frances Baylor Barnum. Novelist
1855	Ernest Chausson. Composer
1857	Lucien Wolf. Historian

1859	Charles Lindbergh, Sr. Congressman
1870	Guillaume Lekeu. Composer
1874	Henry Baird
1875	Frank Hinman Waskey. First congressman from Alaska
1876	Josef Hofman. Pianist
1878	Ruth St. Denis. Dancer
1884	Archbishop John Nutty. Chaplain at West Point
1888	Daniel Callus. Writer
1891	John Bennett. First bishop of Lafayette
1892	Roscoe Otes. Actor
1893	John A. O'Brien. Author
1894	Walter Piston. Composer
1896	George Burns
1903	Leon Ames. Actor
1920	Federico Fillini of film fame
1926	Pat Neal. Actress
1927	Gisele MacKenzie. Actress
1927	Denise McCluggage. Race car driver
1930	Buzz Aldrin. Astronaut
1937	Dorothy Provini. Actress

January 21

1110	St. Amadeus of Lousanne
1337	Charles V. King of France
1685	Susanna Mosier
1734	Sylvanus Cone of Revolutionary War
1743	John Fitch. Steamboat builder
1746	Elkanah Cobb
1812	Edward Tauwitz. Composer
1813	John Fremont. Soldier; explorer; governor
1815	Daniel McCallum of Civil War
1824	Gen. "Stonewall" Jackson. Confederate Field Commander
1825	Valentin Thalhofer. Scholar
1853	Helen Hamilton Gardner of Civil Service Commission
1864	Israel Zangwill. Author
1896	Richard Reid. Editor
1904	Richard Palmer Blackmur. Writer
1909	Bernard Dempsey. Author
1919	Jinx Falkenburg

1922	Paul Scofield. Actor
1926	Steve Reeves. Actor
1926	Telly Savalas
1940	Jack Nicklaus
1941	Placido Domingo. Tenor
1941	Mac Davis. Singer
1947	Jill Eikenberry. Actress

January 22

1440	Ivan the Great
1561	Sir Francis Bacon
1592	Philippe Alegambe. Historian
1631	Vincent Houdry
1690	Nicolas Lancret. Painter
1712	Jonas Langley. Patriot
1729	Gotthold Lessing. Dramatist
1765	Josiah Squire
1788	Lord Byron. Poet
1796	Andrew Lydick. Pioneer
1799	John Hiram Lathrop
1799	Samuel Fullen
1802	Richard Upjohn. Architect
1804	Charles O'Conor
1820	Calvin Goddard. Inventor
1828	Christopher Dice
1832	George Belknap. Founder of U.S. Naval Institute
1857	Gerhard Bente. Editor; writer
1858	Louis Sebastian Walsh. Bishop of Portland; author
1860	Chase Salmon Osborn. Governor of Michigan
1874	Leonard Dickson. Mathematician
1893	Fulton Oursler. Author of *The Greatest Story Ever Told*
1897	Rosa Ponselle. Soprano
1911	Suzanne Danco. Soprano
1919	Jeanette McArthur. Author
1922	Howard Moss. Poet
1935	Pierre du Pont IV. Governor of Delaware
1937	Joseph Wambaugh. Author
1948	George Foreman. World Heavyweight champion
1957	Mike Bossy of ice hockey
1959	Linda Blair. Actress

January 23

1350	St. Vincent Ferrer
1598	François Mansart. Architect
1725	Abraham Littlehole of the Revolutionary War
1730	Joseph Hughes
1737	John Hancock. Signer of Declaration of Independence
1752	Muzio Clementi. Pianist; composer; conductor
1801	Jean Gury
1842	Gen. John Butler of the Battle of Chickamauga
1842	Albert Thompson of the Second Battle of Bull Run
1850	Blandina Segale. Pioneer
1862	David Hilbert. Missionary of the Southwest
1863	Gen. Moses Zalinski
1869	Josiah Flynt. Hobo; author of *Tramping with Tramps*
1875	David Griffith. Director
1876	Bl. Rupert Mayer
1889	Bloodgood Tuttle. Architect
1896	Grand Duchess Charlotte of Luxembourg
1903	Randolph Scott. Actor
1910	Django Reinhardt. Guitarist
1915	Arthur Lewis. Nobel Prize winner in economics
1919	Ernie Kovacs. Actor
1925	Danny Arnold. Writer
1930	William Pogue. Astronaut
1930	Derek Wolcott. Playwright
1933	Chita Rivera. Actress
1953	Pat Haden. National Football League All-Star
1957	Princess Caroline

January 24

76	Hadrian. Roman Emperor
1670	William Congreve. Playwright
1705	Farinello. Castrato soprano. Recipient of the Cross of Calatrava
1709	François Bedos de Celles. Organ authority
1712	Frederick the Great
1733	Gen. Benjamin Lincoln of Continental Army; Secretary of War
1746	Gustav III. King of Sweden
1653	Jacques Alexandre
1760	Matthew Covington

1776	Ernest Hoffman. Poet; composer
1798	Karl Holtei. Poet
1816	Coles Bashford. Governor of Wisconsin
1829	William Mason. Pianist; composer
1850	Mary Murerer. Author
1852	Robert Grant. Writer
1860	Francis Lasance. Author
1862	Edith Wharton. Novelist
1871	Aasar Asche. Actor
1871	Thomas Jagar. Geologist
1872	Morris Travers. Scientist
1879	Gifford Beal. Painter
1889	Vincenzo Davico. Composer
1896	Irving Ives
1915	Robert Motherwell. Painter
1918	Ernest Borgnine
1918	Gottfried von Einem. Composer
1925	Maria Tallchief. Ballerina
1941	Neil Diamond

January 25

1477	Anne of Brittany
1540	St. Edmund Campion. English Jesuit Martyr
1627	Robert Boyle. Scientist
1736	Joseph Lagrange. Inventor of metric system
1746	Thomas Lomax
1759	Robert Burns. Poet
1786	Benjamin Haydon. Painter
1814	Francis Pierpont. Statesman
1817	Mehitable Hill. Settler
1825	Sebastian Buehlhorn. Pioneer
1825	George Pickett. Confederate leader
1854	Mary Shaw. Actress
1860	Charles Curtis. Vice President of the United States
1880	Joseph Bonsirven
1880	Matthew Wall
1889	Francis Talbot. Editor, writer
1895	Florence Mills. Singer
1895	Josephine Pinckney. Singer
1898	Joachim Wach

1903	Adm. Aurelius Vosseller. U-Boat hunter; Under-Secretary of the Navy
1919	Edwin Newman. Writer
1924	Lou Groza of football
1927	Gregg Palmer. Actor
1930	Major Melton. Race car driver

January 26

1518	Giovanni Lanfranco. Developer of baroque style of painting
1678	Joseph I. Holy Roman Emperor
1714	Jean Pigalle. Sculptor
1716	Lord George Germain. British Secretary of State for the American Colonies
1753	Israel Lucas. Patriot
1754	Stephen Lum of the Revolutionary War
1763	Carl XIV Johan. King of Sweden and Norway
1826	Julia Dent Grant. First Lady
1832	Rufus Gilbert. Inventor
1833	Elisabet Ney. Sculptor
1833	Cornelius Bliss. Secretary of the Interior
1839	Anna Hardy. Painter
1842	François Coppee. Poet
1858	Arthur Ingram. Writer
1863	Menzo Goodell
1879	Jenny Marx of the six Marx Sisters
1880	Josephine Brownson. Author
1880	General Douglas MacArthur
1884	Roy Andrews. Explorer
1884	Edward Sapier. Anthropologist
1887	Marc Mitscher. Aviation pioneer
1892	Bede Reynolds. Author of *A Rebel from Riches*
1905	Maria von Trapp. Baroness whose life inspired *The Sound of Music*
1907	Henry Cotton of golf
1911	Polykarp Kusch. Winner of Nobel Prize
1913	William Prince. Actor
1920	Derek Bond. Actor
1929	Jules Feiffer. Cartoonist
1945	Martin Walter. Mathematician

1958   Anita Baker. Singer
1961   Wayne Gretzky of ice hockey

January 27

1571   Abbas I. Shah of Persia
1662   Richard Bentley. Scholar
1745   Elihu Root. Patriot
1745   John Van De Mark of the Revolutionary War
1756   Wolfgang Mozart. Composer
1805   Samuel Palmer. Painter
1814   Eugene Viollet. Architect
1823   Édouard Lalo. Composer
1832   Lewis Carroll. Novelist
1850   Samuel Gompers
1859   Kaiser Wilhelm II
1861   Ralph Modjeski. Designer of Blue Water Bridge
1872   Learned Hand. Federal judge for fifty-two years
1876   Thomas Murphy
1885   Jerome Kern. Composer
1894   Fritz Pollard. Coach
1900   Admiral Rickover
1901   Arthur Rooney. Owner of Pittsburgh Steelers
1920   Frank Albert of the 49ers
1921   Donna Reed
1921   George Mathieu. Painter
1923   Jean Merrill. Writer
1958   Raymond Mauro. Cartoonist

January 28

1382   Richard Beauchamp. Earl of Warwick
1457   Henry VII. King of England
1572   St. Jane Frances de Chantel
1582   John Barcley. Author
1600   Pope Clement IX (Guilio Rospegliosi)
1608   Giovanni Borelli. Scientist
1706   John Baskerville
1726   Laurent Beaumille
1742   William Richards. Patriot

1760	Matthew Carey. Publisher; author
1780	Giovanni Battista Velluti. Last of the castrati soprano
1791	Ferdinand Herold. Composer
1793	Ann Dorr. Settler
1799	Richard Roth. Writer
1809	Richard Whelan. First bishop of Wheeling; founded St. Vincent's College
1827	Colman Seller. Inventor
1848	Commodore William Henry Turner
1849	Denis O'Connell
1873	Gabrielle Colette. Writer
1884	Auguste Piccard. Balloon flyer
1889	William Carrigan. Historian
1890	Harvey Ferguson. Writer
1892	Ernst Lubitsch. Director
1899	Arthur Rubenstein. Pianist
1906	Robert Alton. Actor
1911	George Abbe. Author
1930	Gerhardt Wolf
1932	Parry O'Brien. Track star
1936	Alan Alda. Actor

January 29

1674	Justus Bochmer. Writer
1737	Thomas Paine
1748	William Lovett. Patriot
1754	Moses Cleveland of the Constitutional Convention
1761	Albert Gallatin
1761	Catala Magin. Missionary
1782	Daniel Auber. Composer
1802	John Stewart Barry. Governor of Michigan
1807	Michael Accalti. Missionary
1811	Lorenz Kellner
1843	William McKinley. President of the United States
1846	Frederic Vinton
1862	Frederick Delius. Composer
1862	Clemens Blume
1863	Xavier Gehant
1872	William Rothenstein. Painter
1869	Percival Pollard. Playwright

1878	Barney Oldfield. Race car driver
1879	W.C. Fields
1898	Maria Muller. Soprano
1905	Vina Delmar
1905	Barnett Newman. Painter
1916	Victor Mature
1918	John Forsythe
1923	Paddy Chayefsky. Playwright
1924	Luigi Nono. Musician
1942	Claudine Longet. Singer
1950	Minton Warren. Scholar
1960	Gregory Louganis. Diver

January 30

58BC	Drusilla Livia. Wife of Emperor Augustus; mother of Emperor Tiberius
1756	William Choice. Patriot
1771	George Bass
1775	Walter Landor. Novelist
1789	Wolf Baudissin. Writer
1801	Fr. Pierre Jean De Smet. Missionary
1816	Gen. Nathaniel Banks. Governor of Massachusetts
1817	Wilhelm Wilmer. Philosopher
1823	Capt. George Deshon of West Point; missionary
1841	Guiseppi del Puente. Baritone
1862	Walter Damrasch. Conductor; composer
1868	Helen Dale of Danville
1871	Aaron Tanzer
1882	Franklin Delano Roosevelt. President of the United States
1885	Adm. John Henry Towers. Pacific Fleet Commander
1903	George Hutchinson. Zoologist
1906	Horace Barden. Writer
1911	Roy Eldridge. Jazz musician
1923	Norm Nelson. Race car driver
1930	Dorothy Malone. Actress
1931	Gene Hackman. Actor
1937	Boris Spassky. World champion chess player
1942	Marty Balin. Musician
1945	Martin Adelman. Broadcaster

January 31

- 1597 St. John Francis Regis
- 1640 Samuel Willard. Pioneer
- 1673 St. Louis de Montford
- 1713 Anthony Benezet
- 1748 Isaac Van Aken
- 1752 Gouverneur Morris
- 1769 André Garnerin. First parachutist
- 1812 William Hepburn Russell. A founder of Pony Express
- 1830 James Blaine. Senator; Secretary of State
- 1857 Joseph Braun. Archeologist
- 1872 Zane Grey. Novelist
- 1881 Anna Pavlova. Ballerina
- 1902 Tallulah Bankhead. Actress
- 1905 John O'Hara. Writer
- 1913 Don Hutson of Green Bay Packers
- 1914 Jersey Joe Walcott
- 1915 Garry Moore. Entertainer
- 1916 Frank Parker. Tennis Champion
- 1919 Jackie Robinson
- 1921 John Agar. Actor
- 1923 Joan Dru
- 1929 Jean Simmons. Actress
- 1938 Queen Beatrix of the Netherlands
- 1938 Christopher Burford of Kansas City Chiefs
- 1947 Nolan Ryan

February 1

- 1435 Amadeus IX. Duke of Savoy
- 1462 John Trithemius. Scholar
- 1552 Sir Edward Coke
- 1690 Francesco Veracini. Composer
- 1693 Girolamo Tormiello. Writer
- 1763 Thomas Campbell
- 1800 Elisha Durbin
- 1807 Charles Ignatius White. Editor; writer; publisher
- 1842 Andrew Lambing. Author; historian
- 1844 Alphonse Glorieux. First bishop of Boise
- 1848 Cordelia Howard. Actress

1857	Charles Nordhoff. Writer
1859	Lydia De Witt. Scientist
1859	Victor Herbert. Composer
1860	Edward Cudahy. Established Armour-Cudahy
1872	Clara Butt. Contralto
1872	James S. Daly. Editor; author
1877	Ruth St. Denis. Dancer
1878	Hattie Wyatt Caraway. Senator
1887	Charles Truax
1894	Gardner Hale. Painter
1895	John Ford. Director
1901	Clark Gable
1915	Stanley Matthews of soccer
1918	Muriel Spark. Novelist
1922	Renata Tebaldi. Soprano
1924	Agnes Matthews. Mayor of Wetherfield; state senator

February 2

1208	James I. King of Aragon
1649	Pope Benedict XIII (Pietro Francesco Orsini)
1650	Nell Gwynne. Actress
1711	Prince von Kaunitz-Rietberg
1723	Joshua Cleaves
1735	Uriah Tracy
1757	Igmaco Andio y Varela. Artist
1762	Girolamo Crescentini. Castrato soprano decorated by Napoleon
1797	Hiram Pomeroy
1802	Mancure Robinson. Railroad pioneer
1803	Albert Johnson. Confederate general
1806	Lawrence Hengler. Inventor
1819	Sophie Felix. Actress
1870	Taber Sears. Painter
1875	Fritz Kreisler. Violinist
1882	Prince Andrew of Greece
1882	Geoffrey O'Hara. Composer
1882	James Joyce
1891	Frederick Clifton Grant
1895	George Halas of football
1901	Jascha Heifetz. Violinist
1903	August Beyer. Rural Road Commissioner

1905	Ayn Rand
1906	Gale Gordon
1909	Frank Albertson. Actor
1909	Loren MacIver. Painter
1923	James Dickey. Poet
1946	Howard Bellamy of the Bellamy Brothers
1947	Farrah Fawcett

February 3

1544	Ven. Cesar De Bus
1717	Governor Nicholas Cooke
1736	Johann George Albrechtsberger. Composer
1742	John Worthington Warfield
1750	Benjamin Wiley
1757	Anton Dereser
1795	Antonio Sucre. First president of Bolivia
1803	Albert Johnston. Confederate general killed at the Battle of Shiloh
1808	Alban Stolz. Author
1809	Felix Mendelssohn
1813	Adm. Augustus Chase. Chief of Bureau of Ordnance
1834	Edwin Adams. Comedian
1842	Sidney Lanier. Poet
1851	William Trubner. Artist
1889	Carl Dreyer. Director
1892	John Ewing. Broadcaster
1894	Norman Rockwell
1898	Alvar Aalto. Architect
1900	Mabel Mercer. Singer
1904	Luigi Dallopiceola. Composer
1907	James Michener. Novelist
1926	Glen Tetley
1929	Grace Junker. "Career Woman of the Year 1971-72"
1932	Ivan Davis. Pianist
1940	Fran Tarkenton
1945	Robert Griese of football
1950	Morgan Fairchild
1951	Lund Hargrave. Singer

February 4

- 1575 Pierre de Berulle
- 1688 Pierre Marivoux. Playwright
- 1693 George Lillo. Dramatist
- 1740 Carl Bellman. Poet
- 1762 Randolph Rice
- 1772 Josiah Quincy. Congressman; mayor of Boston; president of Harvard
- 1792 James Gillespie Birney
- 1802 Mark Hopkins
- 1805 William Harrison Ainsworth. Writer
- 1811 Ven. Pierre Eymard
- 1831 Oliver Ames. Governor of Massachusetts
- 1842 George Brandes
- 1881 Fernand Leger. Painter
- 1887 Ludwig Erhard
- 1888 Knute Rockne
- 1892 Ugo Betti. Poet
- 1898 Agnes Ayres. Actress
- 1902 Charles Lindbergh, Jr. Made first flight across Atlantic
- 1904 Clarence Elwell. Bishop of Columbus
- 1906 Clyde Tombaugh. Astronomer; discovered Pluto
- 1912 Erich Leinsdorf. Conductor
- 1913 Dick Seaman. Race driver
- 1913 Woody Hayes
- 1915 Roy Evans
- 1917 Ida Lupino
- 1947 Dan Quayle. Vice President of the United States
- 1950 Pamela Franklin

February 5

- 1723 John Witherspoon. Signer of the Declaration of Independence
- 1725 James Otis. Statesman
- 1730 Constant L'Hommedieu
- 1761 Joseph Dickinson
- 1796 Johannes Geissel
- 1810 Ole Bull. Violinist
- 1827 Sylvester Rosecrans
- 1837 Dwight Moody

1837	Jasper Davidson
1838	Abram Ryan. Poet
1848	Belle Starr. "Bandit Queen" of the Southwest
1868	Maxwell Elliot of the stage
1877	Michael Williams. Writer
1900	Adlai Stevenson II
1902	Floyd Begin. First bishop of Oakland
1906	John Carradine
1914	William Burroughs
1915	Margaret Millar. Writer
1917	Otto Edelman. Singer
1919	Red Buttons
1934	Hank Aaron
1935	Fernando Zawislak. Physicist
1947	Darrell Lee Waltrip. Winner of Winston Cup
1948	Barbara Hershey. Actress

February 6

1478	St. Thomas More. "The King's Good Servant But God's First"; English statesman, executed by Henry VIII for refusal to deny papal supremacy; canonized in 1935
1564	Christopher Marlowe. Poet
1577	Beatrice Cenci
1605	Bl. Bernard of Corleone
1608	Antonio Vieira. Orator
1665	Queen Anne
1756	Aaron Burr. Vice President of the United States
1792	Aghsah Pomeroy
1799	Henry Anderson. Scientist
1814	Fr. Edward Sorin. Founder of Notre Dame, Indiana
1820	Thomas Durant of Union Pacific
1832	John Brown Gordon. Governor of Georgia
1836	Thomas Oliver Selfridge of Civil War; explorer
1838	Henry Irving. Actor
1870	William Auchstetter
1872	Robert Maillart. Bridge builder
1873	Arthur Turner. Conductor
1879	Katharine Fuller Gerauld. Novelist
1895	"Babe" Ruth
1903	Claudio Arrow. Pianist

1904	Harold Gardiner. Author
1904	Ber Dena Belle. Widely acclaimed home economist
1911	Ronald Reagan. Actor; President of the United States
1919	Zsa Zsa Gabor
1933	Mamie Van Doren
1941	Stephen Albert. Composer

February 7

1725	Matheir Rich of the Revolutionary War
1743	Philomon Dorsey
1746	Stephen Tohey. Patriot
1747	Jesse Vaughan
1754	Jacob Lovejoy of the Colonies
1962	Daniel Lucas
1765	Henry Dove
1794	Sidney Morse. Inventor
1802	David Spohr
1809	Frederick Paludin Mueller. Poet
1812	Charles Dickens
1826	Adm. James Jouett. President of Board of Inspection
1830	John Barnhill
1844	Frederick Katzer. Bishop of Green Bay; third archbishop of Milwaukee
1854	Francis Wilson. Actor
1859	Alexander Black. Novelist
1859	Emma Nevada. Opera singer
1867	Laura Wilder. Author of *Little House on the Prairie*
1883	Eubie Blake. Composer
1885	Sinclair Lewis. Novelist
1894	Aaron Hancock. Choctaw minister and missionary
1905	Wally Butts. Coach
1914	Mara Alexander. Actress
1920	Eddie Bracken. Actor
1920	Oscar Brand. Composer

February 8

| 1291 | Afonso IV |
| 1513 | Daniele. Patriarch of Aquileia |

1554	Marina Escobar
1577	Robert Burton. Writer
1649	André Antonil
1649	Gabriel Daniel
1758	Ephraim Rolfe. Patriot
1760	Pleasant Haley of the Revolutionary War
1811	Edwin Dennison Morgan. Union officer; senator; governor
1820	Gen. William Tecumseh Sherman
1828	Jules Verne. Science fiction writer
1838	George Butler. Artist
1851	Kate Chopin
1880	Franz Marc. Painter
1883	Joseph A. Schumpeter. Economist
1888	Edith Evans. Actress
1904	Lyle Talbot. Actor
1906	Harry Roth. Novelist
1906	Chester Carlson. Inventor of Xerox
1906	Adrienne Adams. Writer
1911	Elizabeth Bishop. Poet
1919	Tony Dyer. Actor
1920	Lana Turner. Actress
1925	Jack Lemmon. Actor
1931	James Dean
1942	Robert Klein. Actor
1968	Gary Coleman

February 9

1489	Georg Harlmann. Scientist
1635	Shoreborn Wilson
1698	Joseph Tuttle, Jr. Pioneer
1739	William Bartram. Patriot
1744	Luther Martin. Member of Continental Congress
1773	William Henry Harrison. Ninth President of the United States
1819	William Frith. Painter
1819	Lydia Pinkham
1823	Henry Newton. Inventor
1826	Gen. John Logan of Union Army
1834	Francis Witt. Musician
1836	Franklin Benjamin Gower. Delegate to Pennsylvania Constitutional Convention

1854	Charles Ashburner. Geologist
1863	Anthony Hope. Playwright
1866	George Ode
1874	Amy Lowell. Poet
1891	Ronald Coleman. Actor
1892	Peggy Wood
1909	Heather Angel. Actress
1914	Bill Veeck. Owner of Milwaukee Brewers
1914	Ernest Tubbs
1923	Kathryn Grayson
1927	Tony Maggs. Race car driver
1928	Roger Mudd. Newscaster
1942	Carole King. Singer
1945	Mia Farrow

February 10

1555	Tung Chi Chang. Scholar; painter; writer
1723	James Scott. Patriot
1775	Charles Lamb. Writer
1808	Johan Klotter. Brewer
1808	John Edgar Thomson. President of Pennsylvania Railroad
1810	John Van Buren
1811	John Kerr of Congress
1842	Agnes Mary Clerke. Astronomer
1843	Adelina Patti. Soprano
1846	Ira Remson. President of Johns Hopkins; writer
1868	William Allen White. Editor; writer
1872	John Hartford. President of the Great A & P Tea Company
1893	Jimmy Durante
1887	Charles Sawyer. Secretary of Commerce
1893	Bill Tilden of tennis
1898	Bertolt Brecht. Poet
1898	Judith Anderson. Actress
1906	John Farrow. Director
1910	Maria Cebatari. Soprano
1919	Eddie Johnson. Race car driver
1927	Leontyne Price. Opera singer
1930	Robert Wagner
1934	Fleur Adcock. Author of *Eye of the Hurricane*
1939	Robert Flack
1955	Greg Norman of golf

February 11

- 1535 Pope Gregory XIV (Niccolo Sfondrati)
- 1657 Bernard Fontenelli. Science writer
- 1758 Franz Streher. Numismatist
- 1761 Bartholomew Lot
- 1776 Count Giovanni Capo. First president of Greece
- 1791 Alexandros Mavrokordatos
- 1800 William Talbot. Archeologist
- 1802 Lydia Child
- 1812 Adm. Benjamin Sands. Superintendent of Naval Observatory
- 1820 Theodore O'Hara. Poet
- 1833 Melville Fuller. Chief Justice of U.S. Supreme Court
- 1836 Solomon Washington Gladden
- 1847 Thomas Edison
- 1863 John "Honey Fitz" Fitzgerald. Mayor of Boston
- 1871 Frederic René Coudert
- 1887 Adm. Henry Hewitt. Commander of Allied Forces
- 1900 Thomas Hitchcock of polo
- 1907 William Levitt. Builder of Levittown
- 1909 Max Baer. Heavyweight champion
- 1917 Sidney Sheldon
- 1920 King Farouk
- 1920 Gen. Daniel "Chappy" James
- 1936 Burt Reynolds
- 1938 David Aaker. Writer
- 1953 Philip Anglim. Actor

February 12

- 1368 Sigismund. Holy Roman Emperor
- 1584 Casparus Barlaeus. Historian
- 1739 Job Loring. Patriot
- 1746 Gen. Tadeusz Kosciuszko. Great hero of Poland, aided the colonists fighting in the Revolutionary war; took part in Saratoga and Carolina campaigns
- 1760 Jan Dussek. Composer
- 1768 Frances II. Holy Roman Emperor
- 1775 Louisa Adams (Mrs. John Quincy Adams). First Lady
- 1797 John Timen. First bishop of Buffalo
- 1799 Thankful Weeks. Early settler

1809	Abraham Lincoln. President of the United States
1813	Otto Ludwig. Playwright
1828	George Meredith. Novelist
1829	Gerard Schneeman. Writer
1831	Myra Bradwell
1836	Willis Jackman
1855	Jacob Adler. Actor
1870	Robert Kingsley
1880	John L. Lewis. Labor leader
1884	William Guthrie
1893	General Omar Bradley
1898	Roy Harris. Composer
1903	Andy Harrington of baseball
1904	Ted Mack
1915	Lorne Greene
1917	Al Cervi of basketball
1919	Forrest Tucker
1923	Alan Dugan. Poet
1926	Joan Mitchell. Singer
1927	Anne Gillis. Actress
1934	Bull Russell. Defensive center
1936	Joe Baker. Actor
1952	Henry Rono of track
1960	Cyndi Stoddart. Sculptor

February 13

1440	Hartman Schedel. Historian
1457	Mary, Duchess of Burgundy
1480	Girolamo Aleandro
1533	Christian Adrichem
1599	Pope Alexander VII (Fabio Chigi)
1733	Abel Prindle
1751	John Daball. Patriot
1752	Zephaniah Cudworth
1752	Samuel Bliss. Patriot
1752	Nicholas Brooke of the Revolutionary War
1793	Philipp Veit. Painter
1800	Orange Scott
1805	David Field. Author of the Field Codes
1811	Luigi Tosti. Historian

- 1812 Commodore Samuel Lee
- 1838 Charles Barnard. Author
- 1851 Andrew Gordon. Explorer
- 1851 George Brown Goode
- 1852 Johann Dreyer. Astronomer
- 1864 Edwin Arden. Actor
- 1865 Dugald Jackson. Author
- 1867 James Burns
- 1870 William Roughead. Writer
- 1875 Spencer Nichols. Painter
- 1880 John La Farge
- 1892 Robert Jackson. Supreme Court Justice
- 1902 Waldemar Gurian. Author
- 1904 Georges Cathelot. Tenor
- 1910 William Shackley. Nobel Prize winner
- 1912 Margaretta Scott
- 1918 Pat Berg of golf
- 1918 Oliver Smith. Producer
- 1919 Tennessee Ernie Ford
- 1919 Joan Edwards. Singer
- 1920 Eileen Farrell. Singer

February 14

- 1404 Leon Battista Alberti. Musician
- 1602 Pier Francisco Cavelli. Composer
- 1710 Obadia Olney. Pioneer
- 1731 Winthrop Gilman. Patriot
- 1746 Patrick Hickey. Editor
- 1755 Valentine Balsbaugh. Editor
- 1757 Filding Lewis, Jr.
- 1758 Christopher Longyear of the Revolutionary War
- 1760 Richard Allen.
- 1766 Louis Dubourg. Founder of St. Mary's College
- 1807 Max Ainmiller. Artist
- 1813 Aleksandr Dargomyzhski. Pianist; composer
- 1819 Christopher Latham Shales. Printer; descendant of Priscilla and John Alden
- 1821 Commodore John Fabinger
- 1840 Reginald Buckler. Writer
- 1855 Wallace Yenerick. Settler

1858 Joseph Thomson. Explorer
1882 John Barrymore
1882 Ignaz Friedman. Pianist
1894 Jack Benny
1896 Edward Milne. Scientist
1905 Thelma Ritter
1913 James Hoffa
1921 Jean Demessieux. Organist
1921 Hugh Downs. Broadcaster; consultant; skipper; composer
1930 Monty Manibog
1932 Banjo Matthews. Three time winner of Daytona
1942 Ricardo Rodriquez. Bicycle, motorcycle and auto racing champion
1944 Ronnie Peterson. Race car driver
1946 Gregory Hines. Actor
1960 Jim Kelly of football

February 15

1483 Babur. Founder of the Mogul dynasty
1519 Pedro Mendez. Established Florida; founded the city of St. Augustine
1546 Johann Pistorius. Historian
1571 Michael Praetorius. Composer
1705 Charles von Loo. Painter
1710 Louis XV. King of France
1725 John Calderwood
1734 William Lick. Patriot
1748 Jeremy Bentham of Revolutionary War
1760 Jean François Lesueur. Composer
1803 Johann August Sutter of Sutter's Fort. California pioneer
1812 Charles Tiffany. Jewelry designer
1817 Charles Daubigny. Painter
1825 Nicholaus Haupt
1845 Elisha Root
1855 Hugo Vogel. Painter
1858 Marcella Sembrich. Opera singer
1861 Alfred Whitehead. Mathematician
1863 Francis McNutt. Author
1868 Victor O'Daniel. Historian
1876 Fr. John Farley. Winner of nine letters in football, basketball and track at Notre Dame; long-time rector at Notre Dame

1881   William Sweet. Writer
1892   James Forrestal. Secretary of Navy
1897   Fr. Leonard Feeney. Author; lecturer
1899   Gale Sondergaard. Actress
1905   Harold Arlen. Composer
1907   Cesar Romero. Actor
1931   Claire Bloom. Actress
1935   Roger Chaffee. Astronaut
1948   Ron Cey of baseball

February 16

1355   Bl. Peter of Pisa
1452   Bl. Joanna of Portugal
1698   Pierre Bauger. Inventor
1747   George Schuck
1749   Patience Nickerson
1756   Placidus Braun. Historian
1764   Samuel Jackson of the Revolution
1783   Jean Halloy. Geologist
1804   Jules Janin. Novelist
1812   Henry Wilson. Vice President of the United States
1821   Heinrich Barth. Explorer
1824   Clara Moore. Writer
1826   John Cameron
1829   Sarah Dorsey. Author
1838   Henry Brooks Adams. Writer
1840   Henry Watterson. Journalist; editor, congressman; orator
1841   Nelson Baker. Builder of industrial school, hospital and infants home
1843   Daniel Cook Fisher of the Battle of Winchestor
1876   George Trevelyan. Historian
1887   Adm. Jesse Oldendorf. Battleship leader
1893   Katharine Cornell. Actress
1900   Albert Hacket. Playwright
1903   Edgar Bergen
1907   Edwin Shipstead of Ice Follies
1909   Hugh Beaumont. Actor
1919   Mason Adams. Actor
1930   Jack Sears. Race car driver
1959   Jack McEnroe of tennis

February 17

1588	Jusepe de Ribera. Painter
1653	Arcangelo Carrelli. Violin virtuoso
1729	Obadia Lovell. Patriot
1753	Friedrich Klinger. Novelist
1809	Cyrus McCormick. Developer of reaper
1820	Henri Vieuxtemps. Musician
1827	Maria Rossetti. Writer
1844	Montgomery Ward
1850	Ludwig Bonvin. Composer
1856	Frederic Ives. Inventor
1864	Andrew Paterson. Poet
1877	Pierre Crabites. Writer
1878	Edward Hawks. Author
1879	Dorothy Canfield Fisher. Novelist
1881	Bess Aldrich. Author
1889	H.L. Hunt
1893	Wally Pipp
1895	Adm. Charles Joy. Korean War leader; writer; superintendent of Naval Acadamy
1896	Frank Emmer. Baseball great
1902	Marian Anderson. Singer
1905	Swede Oberlande. All-American
1912	Brusie Ogrodowski of baseball
1919	Todd Bolender
1921	Walter Fairservis. Archeologist
1925	Hal Holbrook
1934	Alan Bates
1936	Jim Brown. Running back
1963	Michael Jordan of Chicago Bulls

February 18

1013	Herman Contractus. Poet
1404	Leone Alberti. Artist
1516	Mary Tudor. Queen of England
1605	Abraham Ecchellensis. Scholar
1657	Denis Le Nourry. Writer
1745	Alessandro Volta. Inventor of electric battery
1796	Daniel Norton. Sailmaker

1805	Louis Goldsborough. Union commander of Civil War
1805	Abram Stratton. Pioneer
1806	Edward Heis. Astronomer
1820	George Vandenff. Actor
1829	Jack Ambler of Congress
1833	Alfred Irion. Legislator
1842	Charles Emory Smith. Postmaster General
1860	Anders Zorn. Sculptor
1889	Cardinal Aloisius Muench. First American of the Curia
1890	Edward Arnold. Actor
1892	Wendell Wilkie
1898	Enzio Farrari of racing
1908	Dee Brown. Author of *Bury My Heart at Wounded Knee*
1914	Pee Wee King. Singer
1915	Dane Clark. Actor
1923	Peggy Mensinger. Mayor of Modesto
1926	George Kennedy. Actor
1950	Cybill Shepherd. Actress
1952	Juice Newton
1954	John Travolta
1957	Marita Koch. Sprinter
1964	Matt Dillon

February 19

1652	Dorothy Bushnell
1704	Jean Baptiste Lemoyne. Artist
1706	John Hornyold. Settler
1717	David Garrick. Patriot
1743	Luigi Boccherini. Composer
1749	Laurence Litchfield of the Revolution
1766	William Dunlap. Artist
1789	Sir William Fairbairn
1792	Roger Murchison
1793	Sidney Rigdon
1828	Delos Ashley. Congressman
1866	Thomas Jefferson Jackson See. Astronomer
1877	Gabriel Munter. Artist
1895	Louis Cahern. Actor
1899	Lucio Fontana. Artist
1902	John Bubbles. Dancer

1903   Kay Boyle. Author
1911   Merle Oberon
1916   Eddie Arcaro. Jockey
1917   Carson McCullers. Author of *The Heart is a Lonely Hunter*
1924   Lee Marvin. Actor
1926   Ross Thomas. Mystery writer
1927   Ernest Trova. Sculptor
1932   Joseph Kerwin. Astronaut
1940   Smoky Robinson
1944   Christine Reed. Mayor of Santa Monica
1954   Frank Hill. Legislator
1960   Prince Andrew

February 20

1469   James de Vio
1706   Claude Coquart. Missionary
1719   Joseph Bellamy of the Colonies
1736   Maj. John Bradford. First deputy to court of Massachusetts
1743   Daniel Lowell. Patriot
1772   Isaac Chauncey of the Navy
1778   Margaret Smith. Author
1791   Karl Czerny. Composer
1816   William Rimmer. Sculptor
1817   Catherine Forrest. Actress
1825   William Allen Butler. Poet
1829   Joseph Jefferson. Actor
1872   Charles Trumbull. Editor
1877   Mary Gordon. Soprano
1888   Marie Rambert. Ballerina
1888   Georges Bernanos. Writer
1892   Raymond Moore. Geologist
1897   Ivan Albright. Painter
1902   Ansel Adams. Photography pioneer
1906   John J. Swalve. Mathematician
1907   Malcomb Atterburg. Actor
1924   Gloria Vanderbilt
1926   Bob Richards. Pole Vaulter
1927   Hubert de Givenchy. Fashion designer
1927   Roy Cohn
1931   Amanda Blake

1934	Bobby Unser. Winner of Pike's Peak Hillclimb
1937	Roger Penske. Race car driver
1946	Sandy Duncan
1948	Jennifer O'Neill

February 21

1683	Johann Christoph Wolf. Historian
1684	Mary Butler. Early settler
1723	Louis Anquetil. Historian
1742	Enoch Levering. Patriot
1755	Anne Grant. Author
1777	Abel Huntington of Congress
1791	John Mercer. Developed mercerization of cotton fabric
1779	Friedrich Savigny. Scholar
1794	Gen. Antonio López de Santa Anna. Captured the Alamo
1801	John Henry Cardinal Newman
1810	Carl Stohlman
1817	Josa Moral. Poet
1821	Charles Scribner. Publisher
1822	Mother Mary of St. Angela Gillespie. Foundress of the Sisters of the Holy Cross in the United States; established hospitals to care for wounded in Civil War
1822	Robert Shufeldt. President, Naval Advisory Board; author
1836	Leo Delibes. Composer
1867	William Faber
1867	Elizabeth Phelps Jordan
1876	Constantin Brancusi. Sculptor
1877	Reginald Garrigau-Lagrange
1889	Felix Aylmer. Actor
1893	Andres Segovia. Guitarist
1895	Carl Dam. Discovered Vitamin K
1899	Bernard Griffin
1903	Raymond Queneau. Poet
1913	Glenn Anderson. Congressman
1925	Sam Peckinpah. Filmmaker
1927	Erma Bombeck
1947	Tyne Daly. Actress
1961	Christopher Atkins. Actor

February 22

- 1402 Charles VII of France
- 1698 St. John Baptist de Rossi
- 1732 George Washington
- 1749 John Lund
- 1752 Patrick Zimmer. Philosopher
- 1787 Commodore George Washington Rodgers. Served on the "USS Constitution"; awarded Congressional Medal at end of War of 1812
- 1793 Isaak Jost. Historian
- 1794 Joseph Duncan. Governor of Illinois
- 1796 Lambert Quetelet. Astronomer
- 1800 William Barnes. Poet
- 1817 Niels Gade. Composer
- 1819 James Russell Lowell. Poet
- 1822 Frances Barraro. Writer
- 1845 Charles Widor. Composer
- 1855 Henry Gann. Composer
- 1858 Lady Frances Balfour. Author
- 1862 Connie Mack of baseball
- 1883 Marguerite Clark. Actress
- 1885 George Finnegan. Army chaplain; bishop of Helene
- 1892 David Dubinsky
- 1904 Peter Hurd. Painter
- 1917 Jane Auer Bowles. Writer
- 1918 Sid Abel of hockey
- 1927 Guy Mitchell. Singer
- 1934 Sparky Anderson of baseball
- 1937 Tommy Aaron of pro golf
- 1943 Dick Van Arsdale. Broadcaster
- 1949 Niki Lauda. Winner of Grand Prix
- 1950 Julius Erving of basketball
- 1975 Drew Barrymore

February 23

- 1417 Pope Paul II (Pietro Barbo)
- 1440 Matthias Corvinus. King of Hungary
- 1526 Gancalo Silveira. Missionary
- 1680 Jean Bienville. Explorer

1685	George Frideric Handel. Composer of "Messiah"
1723	Richard Price. Philosopher
1731	Increase Crosby
1734	Elizur Goodrich
1739	Jean Hubert
1750	John Custis
1751	Henry Dearborn. Secretary of War
1813	John Murray Forbes of railroad fame
1822	Giovanni De Rossi. Archeologist
1823	James Batterson. Founder of Travelers Insurance
1828	Edward Goodfellow. Editor
1842	John Brondel. Settler
1843	Lazette West. Pioneer
1857	Margaret Deland
1861	Frederick Burton. Composer
1863	Franz von Stuck. Painter
1881	Titus Brandsma. Martyr
1887	Jackson Adair of Congress
1900	Hermann Buckers. Scholar
1904	William Shirer. Author
1928	Hans Hermann. Winner of Rhineland Cup Sports Car Race
1934	Charles Barnett. Writer
1939	Takako Asakawa. Dancer
1943	Lee Allison. Director
1948	Peter Anastos. Choreographer

February 24

1500	Charles V. Holy Roman Emperor
1540	Don Juan of Austria. Hero of the Battle of Lepanto
1557	Matthias. Holy Roman Emperor
1586	M. Faber. Writer
1595	Matthias Sarbiewski. Poet
1697	Bernard Albinus
1739	Michael Blessing
1758	John Cabbage
1771	Johann Cramer. Composer
1827	Gen. Elisha Bassett Langdon of the Civil War
1833	Eduard Taaffe. Premier of Austria
1836	Winslow Homer. Painter
1840	Jean Berthier. Missionary

1842	Arrigo Bacto. Composer
1844	William Russell. Novelist
1862	Franz Clemens. Philosopher
1865	John Garvin. Author
1867	James Anthony Welsh. Author
1869	John Buschemeyer. Mayor of Louisville
1874	John Wagner of baseball
1885	Adm. Chester Nimitz. Commander-in-chief of Pacific Fleet
1890	Marjorie Main. Actress
1893	Fern Andra. Actress
1908	Audrey McDaniel. Author
1931	James Abourezk. Congressman
1935	Renata Scatto. Soprano
1936	Lance Reventlow. Race car driver: winner of Nassau Cup

February 25

1304	Ibn Battuta. Fourteenth century traveler
1647	Rebecca Chittenden
1746	Abel Catlin
1788	Sarah Olney of Providence
1811	Joseph Tuthill of Congress
1816	Parke Godwin. Editor
1829	Odo William Russell. Ambassador
1841	Pierre Auguste Renoir. Painter
1842	Ida Lewis. Lighthouse keeper
1846	Denis Bradley. Pioneer
1848	Eugene Melchior. Writer
1848	George Turner. Senator
1848	Edward Harriman. President of Sodus Bay & Southern Railroad
1859	Jack Burke. Governor of North Dakota
1859	William Zimmerman. Architect
1871	John Tucker. Author
1872	James Gordon. Missionary
1873	Enrico Caruso
1904	Adelle Davis
1905	Perry Miller. Writer
1908	Frank Slaughter. Novelist
1913	Jim Backus. Actor
1917	Anthony Burgess. Writer
1932	Faron Young. Singer

1938   Herb Elliott. Runner; world record holder
1943   George Harrison. Composer

February 26

1361   Wenceslaus IV. King of Bohemia
1514   Otto of Augsburg
1672   Augustin Colmet
1747   Johannes Schlencker. Patriot
1748   Comfort Sands of the Revolution
1759   Ludwig Jakob. Scholar
1774   Pierre Rode. Composer
1788   John Reynolds. Governor
1800   John Baptist Purcell
1802   Victor Hugo
1808   Honore Daumier. Painter
1814   Charles Sainte-Claire Deville. Geologist
1818   Emperor Alexander III
1830   Morgan Jones of Congress
1832   John Nicolay. Author
1839   Xanthus Russell Smith. Painter
1840   Eugene Schuyler. Author
1846   "Buffalo Bill" Cody
1866   Herbert Dow
1874   Arthur Stringer. Author
1879   Frank Bridge. Musician
1887   Pete Alexander of baseball
1906   Madeleine Carroll. Actress
1914   Robert Alda
1916   Jackie Gleason
1917   Ming Pei. Architect
1919   Mason Adams. Actor
1920   Tony Randall
1922   Margaret Leighton. Actress
1933   Godfrey Cambridge

February 27

1606   Laurent de La Hire. Painter
1635   Francesco Baldovin. Poet

1714	Benhard Havestadt. Settler
1799	Frederick Catherwood. Architect
1807	Henry Wadsworth Longfellow
1810	John Gilbert. Actor
1823	William Buel Franklin. Commander at Bull Run, Malvern Hill and Fredericksburg
1831	Adolph Beer. Historian
1836	Russell Alger. Senator, governor; Secretary of the Navy
1837	Francesca Alexander. Artist
1847	Ellen Terry. Actress
1850	Henry E. Huntington. Founder of Huntington Library
1855	Mary Davies. Mezzo soprano
1856	Mattia Battistini. Baritone
1876	Edward Garesche. Editor
1893	Issai Delrowen. Conductor
1896	Adm. Arthur William Radford. Pacific Fleet Commander
1897	Bernard Lyot. Astronomer
1902	Marian Anderson. Contralto
1902	John Steinbeck. Author
1902	Gene Sarazen. Winner of U.S. Open
1910	Joan Bennet. Actress
1910	Peter de Vries. Author
1913	Irwin Shaw
1926	Eligio Cremonina. Cellist
1934	Ralph Nader
1935	Mirella Freny. Soprano
1939	Peter Revson. Race car driver
1939	Antoinette Schey. Ballerina
1943	Rosemary Mayor. Artist

February 28

1533	Michel de Montiegne. Writer
1630	Mathias Tanner
1675	Guillaume Delisler. Cartographer
1690	Alexis of Russia
1704	Louis Godin. Astronomer
1743	Rene Havy. Chrystallagrapher
1762	Jonathan Large of the Revolution
1776	Jean Pierre Boyer. President of Haiti
1783	Gabriella Ressetti. Poet

1797	Edmund O'Callahan. Historian
1814	William Kingston. Novelist
1838	Phillippe Villiers. Writer
1854	William Cockrow. Orator
1869	William Veacie Pratt. President of Naval War College
1871	Isabel Irving
1872	Walter Willard Abell. Publisher
1877	Henri Brevil. Scientist
1893	Ben Hecht. Writer
1895	Marcel Pagnol. Dramatist
1905	Joseph Breig. Author
1910	Vincent Minelli. Director
1912	Michael Walsh. President of Boston College and Fordham University
1915	Zero Mostel
1923	Charles Durning. Actor
1928	Betty Ackerman. Actress
1940	Mario Andretti. Race car driver
1941	Nathaniel White. Astronomer
1954	Nancy Allison. Dancer

February 29

1468	Pope Paul III (Alessandro Farnese). Launched the Council of Trent
1472	Bl. Antonio of Florence
1707	Col. Peter Jefferson. Father of Thomas Jefferson
1728	Dr. Daniel Cresap. Patriot
1728	Robert Gage. Novelist
1736	Stephen Butterfield of the Revolutionary War
1736	Ann Lee Standerin. Brought the Shaking Quakers to America
1747	Jacob Cushman. Patriot
1748	Peter Ackerman. Fought for the Colonies
1748	Henry Biles. Patriot
1792	Karl Baer. Scientist
1792	Gioaschino Rossini. Composer
1820	Lewis Albert Sayre. Inventor
1824	Mary Patch of Charlestown
1824	John Avery. Congressman
1824	Washington Bartlett. Governor of California
1828	Emmalina Wells. Editor; poet; founder of Utah Women's Press Club

1840	John Holland of submarine fame
1844	Adm. French Ensor Chadwick. Hero of Spanish-American War
1844	Adm. Colby Chester. Superintendent of U.S. Naval Observatory
1852	Mark Reeves Bacon. Congressman
1860	Herman Hollerith. Computer pioneer
1860	Ann Gugerty Shehan. Settler
1880	Paul Blakely. Journalist
1896	Ralph Miller of baseball
1904	Jimmy Dorsey. Bandleader
1916	Frank Charles Arrance. Scientist
1920	Michelle Morgan
1924	Al Rosen of American League
1928	Jack Ackland. Actor
1936	Sandra Leitzinger. Artist
1936	Alex Rocco
1940	Bishop John Ricard
1960	Karen Chapman. Actress

March 1

1647	Bl. John de Britto. Martyr
1649	Ven. Angela Maria of the Immaculate Conception
1721	Israel Standish
1749	Seth Church
1764	Jeptha Lee
1783	Gabriel Sossetti. Poet
1795	Joseph Damberger. Historian
1807	Blakely Brush
1810	Fryderyk Chopin. Composer
1811	Leander Babcock of Congress
1812	Latimer Ballow
1817	Giovanni Dupre. Sculptor
1832	Frederic Coudert. Orator
1833	Bernard Jungman. Historian
1841	Blanche Bruce. Senator
1846	William Demal. Composer
1848	Abner Tannenbaum. Journalist
1854	Amelia Schertz. Pioneer
1855	Mary Ella Waller. Novelist
1861	Henry Harland. Novelist
1871	Max Goldschmidt. Cantor of Niederweisel

1877	Albert Dupois. Composer
1882	Gaston Lachaise. Sculptor
1885	Lionel Atwell. Actor
1886	Osker Kokoschka. Writer
1887	Percival Wilde
1896	Dimitri Mitropaulus. Conductor; composer; pianist
1904	Glenn Miller. Bandleader
1909	David Niven. Actor
1917	Cliffie Stone. Singer
1921	Terence Cardinal Cooke. Archbishop of New York; military vicar of the armed forces
1921	Dinah Shore
1921	Richard Wilbur. Poet
1924	Deke Slayton. Astronaut
1926	Pete Rozelle. Commissioner of National Football League
1927	George Abell. Astronomer
1927	Harry Belafonte
1927	Lucine Amara. Singer
1944	Roger Daltrey. Singer

March 2

1315	Robert II. King of Scots
1459	Pope Adrian VI (Adrian Florensz)
1728	Gideon Brainard
1736	Michael Desgranges. Patriot
1752	Jacob Bittenbender of the Revolutionary War
1757	Wait Chatterton
1769	De Witt Clinton. Statesman
1778	William Austin. Writer
1779	Joel Poinsett. Secretary of War
1793	Sam Houston. Commander-in-chief of Texas Army which defeated Santa Ana; U.S. Senator; governor; president of the Republic of Texas
1810	Pope Leo XIII (Gioacchino Pecci). Author of encyclical *Rerum Novarum* on the conditions of workers in the modern world
1819	Sarah Royce. Pioneer
1819	Samuel Brennan. Inventor
1821	George Banks. Composer
1824	Gen. Henry Beebee Carrington
1824	Bedřich Smetana. Composer

1829	Carl Schurz. Union general; Secretary of the Interior
1834	Edward Joyner
1869	Albert Kahn. Architect
1873	Inez Irwin. Author
1876	Pope Pius XII (Eugenio Pacelli)
1898	Mark Hines Harris. President of Tennessee Baptist Convention
1904	Dr. Seuss. Writer
1917	Desi Arnaz
1917	Jim Konstanty of baseball
1919	Jennifer Jones
1921	Brenda Lewis
1930	Stephen Sondheim. Composer
1934	"Hopalong" Cassidy of football
1942	John Irving. Author

March 3

1500	Cardinal Reginald Pole. Last Catholic Archbishop of Canterbury
1737	Benjamin Cooley. Patriot
1756	William Goodwin. Writer
1760	José Senan. Missionary
1780	José Zalvidea. Missionary
1781	George Mitchell of War of 1812
1793	William McReady. Actor
1822	James Nugent
1824	Adm. Edward Simpson
1825	Annie Keary. Novelist
1831	George Pullman. Developer of sleeping car
1835	John Janssen. Missionary: first bishop of Belleville
1840	Constance Woolson. Writer
1844	François Del Mal
1845	George Cantor. Discovered transfinite numbers
1847	Alexander Graham Bell
1848	Lilian Nielson. Actress
1856	Max Uhle. Scholar; explorer
1857	Alfred Bruneau. Composer
1869	Julia Arthur. Actress
1873	William Green. Labor Leader
1875	Robert Frost. Poet
1895	Gen. Matthew Ridgeway
1901	Charles Goren of bridge

1911 Jean Harlowe
1962 Jaqueline Joyner Kersee. Gold Medal winner

March 4

1394 Henry the Navigator
1678 Antonio Vivaldi. Violinist
1720 Joseph Crocker of the Revolutionary War
1748 General Casimir Pulaski. Served at Brandywine, Trenton and Germantown
1751 Anthony Cuthbert of the Colonies
1753 Solomon Cook. Patriot
1756 Henry Raeburn. Painter
1773 Frances Slocum. Pioneer
1792 Isaac Lea. Writer
1794 William Carleton. Author
1812 Levin Reichel. Historian
1812 William Morton Reynolds. Editor
1818 Tobias Mullen. Missionary
1826 Theodore Judah. Railroad pioneer
1835 Giovanni Virginio Schiaparelli. Astronomer
1861 Arthur McGiffert
1864 Adm. David Taylor. Pioneer of modern warship design
1880 Channing Pollock. Playwright
1888 Knute Rockne. Coach of Notre Dame
1895 Milt Gross. Cartoonist
1897 Lefty O'Doul of baseball
1921 Joan Greenwood. Actress
1933 Nino Vaccarella. Race car driver
1939 Joanne Carner. Pro golfer

March 5

1133 Henry II. First Plantagenet King of England
1324 David II. King of Scotland
1326 Lovis I. King of Hungary
1512 Gerardus Mercotor. Cartographer
1658 Antoine Cadillac. Founder of Detroit; governor of Louisiana
1696 Giovanni Battista Tiupolo. Venetian painter
1732 Ven. Domenico Blasucci

1761	Joseph Dickinson
1789	William Archer of Congress
1794	Robert Grier. Supreme Court Justice
1819	Anna Mowatt. Actress
1821	Anne Demeuer. Soprano
1824	James Ives
1843	Miles Davis. Author
1851	Herman Ridder. Publisher
1852	Isabella Gregory. Founder of Abbey Theatre
1864	Louise Miln. Novelist
1870	Frank Norris. Novelist
1885	Audley Anderson
1887	Heiter Villa. Composer
1789	Michael Ash. Congressman
1903	James Beard
1908	Rex Harrison
1924	Ken Tyrell. Race car driver
1929	Joan Shawlee. Actress
1939	Samantha Eggers

March 6

1475	Michelangelo
1508	Humayun. Emperor of India
1591	Thomas Tamburini
1640	Marc Antonio Barbarigo
1724	Henry Laurens
1754	John Linderberger
1776	Luigi Lambruschini
1786	Adm. Daniel Patterson. Naval Commander of War of 1812
1786	Sir Charles Napier. Commander of the Crimean War
1803	Charles Acton
1806	Elizabeth Barrett Browning. Poet
1823	Marietta Alboni. Singer
1826	Richard Henry Alvey. Called for Maryland's secession from the Union
1831	Gen. Philip Henry Sherman of the Civil War
1835	Gen. Charles Ewing of the Civil War
1840	Gilderay Griffin. Author
1858	Gustav Wied. Novelist
1885	Ring Lardner

1886   Bill Sweeney of Chicago Bears and Boston Braves
1892   Emmet Walsh. Bishop of Charleston
1896   Adm. William Fechteler. Commander-in-chief of Allied Forces in Southern Europe
1900   Leo Dworschak. Bishop of Fargo
1900   Gina Cigna. Soprano
1903   Empress Nagako of Japan
1905   Bob Wills. Fiddler and singer
1908   Lou Costello. Comedian
1918   Jo Mary McCormick-Sakurai. Writer
1924   Sarah Caldwell. Conductor; director

March 7

1536   Achille de Harley
1543   Bl. John Larke. Martyr
1564   Pierre Coton
1693   Pope Clement XIII (Carlo Rezzonico)
1707   Stephen Hopkins. Chief Justice of Rhode Island; signer of the Declaration of Independence
1725   Josia Fassett. Settler
1726   Abner Sampson. Pioneer
1733   Richard Law. Judge; first mayor of New London
1733   Benjamin Raymond. Patriot
1743   Joseph Dorsett of the Colonies
1743   Thomas Clowes. Early American settler
1743   Luther Whipple. Patriot
1749   George Diel of Revolutionary War
1757   James Potts
1760   Elijah Boardman of Colonel Charles Webb's regiment
1770   Gregorius Ziegler
1785   Alessandro Manzoni. Novelist
1820   Francis Wharton. Writer
1829   Eduard Vogel. Explorer
1832   Orlando Poe
1837   Henry Draper. Astronomer
1841   William Rockhill Nelson. Journalist
1845   Daniel David Palmer. Founder of Chiropractic
1868   Sewell Ford. Writer
1872   Piet Monrian. Painter
1875   Maurice Ravel. Composer

1875	Mary Teresa Norton of Congress
1886	Moorhouse Miller. Editor; writer
1913	Margaret Clark. Author
1930	Lord Snowden
1938	Janet Guthrie. Race car driver
1944	Elton Gallegly. Congressman
1960	Joe Carter of baseball

March 8

1495	Giovannie Rosso. Painter
1495	St. John of God
1714	Carl Bach. Composer
1726	Richard Howe. British admiral
1729	Samuel Bradstreet of the Colonies
1750	Ebenezer Buck. Patriot
1783	Hannah Van Buren (Mrs. Mactin Van Buren)
1822	Richard Johnston. Author
1824	Frederic Giborne. Inventor
1831	Sam Jaffe. Actor
1840	Franco Faccio. Conductor; composer
1841	Oliver Wendell Holmes, Jr. of Supreme Court
1856	Mary Wright Plummer
1858	Ruggiero Leoncavallo. Composer
1876	Franco Alfano. Composer
1879	Otto Hahn. Recipient of Nobel Prize
1886	Edward Kendall. Scientist
1888	Stuart Chase
1890	Gene Fowler. Writer
1900	Howard Aiken. Inventor
1902	Jennings Randolph. Congressman
1939	Lynn Seymouir. Ballerina
1942	Sheila Allen. Writer

March 9

1564	David Fabricius. Astronomer
1568	St. Aloysius Gonzaga. A model of holy purity
1631	Claude Menestrier. Antiquarian
1697	Karoline Neuber. Actress

1734	Elisha Hinman. Captain of Revolutionary War
1742	Samuel Provoost. Patriot
1753	Jaques Champeaux of the Revolution
1763	William Cobbett. Author
1773	Isaac Hull. Commander of the "Constitution" during War of 1812
1789	John Harney. Poet
1806	Edwin Forrest. Actor
1815	David Davis
1824	Leland Stanford. President of Central Pacific Railroad founder of Stanford University; senator; governor
1856	Eddie Foy
1868	Charles Warren. Historian
1881	Ernest Bevin
1902	Will Greer. Actor
1903	Albert Gregory Cardinal Meyer. Bishop of Superior; archbishop of Milwaukee; cardinal-archbishop of Chicago
1905	Rex Warner. Author
1906	David Smith. Sculptor
1910	Samuel Barker. Composer
1918	Mickey Spillane. Author
1921	Carl Betz. Actor
1926	Neil Armstrong of Philadelphia Eagles
1930	Thomas Schippers. Conductor
1937	Brian Redman. Auto racer
1943	Bobby Fischer. World chess champion

March 10

1280	Piers de Mauley, Lord of Mulgrave Castle
1452	Ferdinand II. King of Aragon
1653	John Benbow
1675	Mary Bushnell. Early settler
1726	Nathaniel Abbott. Patriot
1734	Seth Currier of the Colonies
1749	Lorenzo Da Ponte. Poet
1749	John Playfair. Mathematician
1772	Frederick Schlegel. Poet
1777	Robert Allison. Congressman
1805	Heinrich Roedter. Ohio legislature; editor
1817	Claude Dubois. Missionary

1821 Adm. James Nicholson of Civil War
1822 Rebecca Lance
1832 Johann Bellermann. Composer
1841 Ina Coolbrith. Poet
1862 Emily Johnson. Writer
1892 Arthur Honneger. Composer
1900 Sherman Billingsley. Owner of the Stork Club
1905 Victor Anfuso of Congress
1911 Warner Anderson. Actor
1917 David Hare. Sculptor
1938 Norman Blake. Singer
1939 Ann Atmore
1940 Chuck Norris. Actor
1941 Sandra Palmer. Golfer

March 11

1544 Torquato Tasso. Poet
1681 Richard Godfrey of Queen Anne's War of 1702
1725 Henry Benedict Stuart. Cardinal of York; titular king of Great Britain and Ireland
1728 Barnabas Bangs. Settler
1748 Daniel Alderman
1753 John Sweet. Patriot
1758 George Bark of the Colonies
1759 Aaron Ames of the Revolutionary War
1760 George Bell. Patriot
1762 Annie Bradbury. Settler
1796 Francis Wayland
1808 Peter Bell. Governor of Texas
1819 Marius Petipa. Dancer
1834 George Ramsdell. Governor of New Hampshire
1839 Arthur Gorman. Senator
1840 Kezia McGinnis
1843 Frank Murdock. Actor
1843 Sydney Fenn Smith. Journalist
1876 Carl Ruggles. Composer
1893 Wanda Gag. Illustrator
1897 Eva Betz. Writer
1898 Dorothy Gish. Actress
1933 Perseus Adams. Poet

1936 Antonin Scalia. Supreme Court Justice
1943 Arturo Merzario. Racing driver

March 12

1270 Charles of Valois
1586 Jean Delbeau. Missionary to Canada
1626 John Aubrey. Writer
1700 Filippo Acciaioli
1703 Resolved Waterman. Early settler
1710 Thomas Arne. Composer
1747 John Crosby. Patriot
1749 Preserved Dakin of the Colonies
1758 Jesse Lee. Chaplain of Congress and Senate
1763 Benjamin Broch
1790 John Daniell. Inventor
1796 Thomas Reynolds. Governor
1806 Jane Pierce. First Lady
1809 Benjamin Barrett
1812 John Murphy. Publisher
1822 Thomas Buchanan Read. Artist
1835 Simon Newcomb. Astronomer
1844 Solon Brown
1845 Blanch Whiffen. Actress
1862 Jane Delano. Pioneer in combatting yellow and scarlet fever
1882 Commodore John J. Glasson. Commander of the "Lexington"
1877 Annette Abbott Adams. Justice
1873 Stewart White. Author
1878 St. Gemma Galgani
1882 Erwin "Cannon Ball" Baker. Race car driver
1906 John Arledge. Actor
1914 Tommy Farr. Boxer
1921 Gordon MacRae. Actor
1923 Walter Schirra. Astronaut
1938 Johnny Rutherford. Race car driver
1942 Bert Companeris of major leagues
1946 Liza Minelli

March 13

1372 Louis Orleans
1599 St. John Berchmans

1615	Pope Innocent XII (Antonio Pignatelli)
1673	Bridget Hoar
1741	Joseph II. Holy Roman Emperor
1756	Phineas Scott. Patriot
1764	Charles Grey. Statesman
1779	Oliver Shaw. Composer
1806	Adolphe Didron. Archeologist
1855	Percival Lowell. Astronomer
1860	Hugo Wolf. Composer
1870	William Glackens. Artist
1884	Hugh Walpole. Novelist
1890	Fritz Busch. Conductor
1897	Paavo Nurmi of Olympics
1899	J.H. Von Vleck. Winner of Nobel Prize
1902	Joseph McGucken. Archbishop of San Francisco
1913	Sammy Kaye. Bandleader
1913	William Casey. Undersecretary of State; headed Security and Exchange Commission; Director of CIA
1916	Ira Hutton. Bandleader
1917	Joe Walsh of baseball
1930	Liz Anderson. Singer
1932	Jan Howard. Singer
1933	William Creser. Composer
1939	Neil Sedaka

March 14

1593	George de La Tour. Artist
1674	James Taylor. Colonel of Militia; member House of Burgesses; Surveyor General
1681	George Telemann. Composer
1754	Charles Reno. Patriot
1781	Jonathan Hunton. Governor of Maine
1797	Asa Whitney. Railroad Promoter
1808	Narcissa Whitman
1820	Victor Emmanuel II. King of Italy
1823	Theodore Banville. Poet
1844	Humbert I. King of Italy
1854	Thomas Marshall. Vice President of the United States
1861	Richard Eugene Burton. Poet
1864	Casey Jones of railroad fame

1871 Olive Fremstad. Mezzo-soprano
1875 Patrick McLane. Congressman
1877 Edna Chase. Editor of *Vogue*
1882 Thomas Alva Yon of Congress
1885 Bruno Hagspiel. Editor
1915 Alexander Bratt. Musician
1927 Bill Rexford. Race car driver
1933 Frank Borman. Astronaut
1933 Michael Caine. Actor
1933 Quincy Jones. Conductor; composer
1934 Eugene Cernan. Astronaut

March 15

1455 Pietro Accolti. Archbishop of Ravenna
1591 Alexander Rhodes. Author
1713 Nicolas de Lacaille. Astronomer
1743 Nicholas Cruger. Patriot
1767 Andrew Jackson. Seventh President of the United States
1797 Brandes Fisher
1800 James Hacket. Actor
1809 Karl Hefele
1821 Gen. Peter Sullivan of Civil War
1823 Albert Stockl
1831 Daniel Comboni
1848 Dame Madge Kendall. Shakespearean actress
1852 Isabella Gregory. Playwright
1856 John Farrelly. Bishop of Cleveland
1875 Lee Scubert
1887 Francis Matthews. Secretary of the Navy
1893 Thyra Winslow. Writer
1893 James H. Van Der Veldt. Scholar; author
1902 Richard Daley. Mayor of Chicago
1902 Wolcott Gibbs. Writer
1903 Abubakar III
1905 Margaret Webster. Director
1913 MacDonald Carey. Actor
1921 Maureen Daly. Novelist
1926 Norm Van Broklin of Rams
1932 Alan Bean. Astronaut
1933 Cecil Taylor. Pianist; composer
1935 Judd Hirsch. Actor

March 16

1581	Pieter Hooft. Poet
1585	Gerbrand Bredero. Playwright
1638	François Crepieul
1695	William Greene. Governor of Rhode Island
1739	George Clymer. Signer of the Declaration of Independence and of the U.S. Constitution
1743	Pardon Daval of the Revolutionary War
1751	Joel Fosgate. Patriot
1751	James Madison of Continental Congress; fourth President of the United States
1763	Mary Berry. Writer
1822	Rosa Bonheur. Artist
1829	Charles Goodrich
1863	Willis Abbot. Historian
1877	Thomas Wyatt Turner
1878	Clemens August Cardinal Galen. Bishop of Muenster, Germany, noted for his public opposition to nazism
1889	Peter Dunne. Historian
1890	Fritz Busch. Conductor
1891	Irita Van Doren. Director of *New York Herald Tribune Book Review*
1896	William Langer. Historian
1900	Cyril Hume
1903	Victor Weybright. Writer
1912	Pat Nixon. First Lady
1916	Lloyd McBride. Labor leader
1920	John Addison. Composer
1925	Marion Montgomery. Writer
1927	Olga San Juan. Actor
1928	Christa Ludwig. Mezzo-soprano

March 17

1473	James IV. King of Scotland
1551	Martin Delrio. Scholar
1578	Francesco Albani. Artist
1628	François Girardon. Sculptor
1720	Elijah Cady. Patriot
1728	Jacobus Blouvelt

1750   Josiah Arms of the Revolutionary War
1777   Roger Brooke Taney. Chief Justice of Supreme Court
1798   Abigail Fillmore. First Lady
1811   Patrick Donahoe. Publisher
1820   Anne Bronte. Writer
1821   Adelia Cleopatra Graves. Author
1834   Gottlieb Daimler. Pioneer in development of internal combustion engine
1839   Joseph Rheinsberger. Composer
1840   Henri Didon. Writer
1846   Kate Greenaway. Artist
1847   Clara Morris. Actress
1856   William George Bruce. Founder of *American School Journal*
1860   Anna Jameson. Writer
1866   Pierce Butler. State senator; Supreme Court Justice
1884   Joseph Bonnet. Organist
1887   Wilfrid Parsons. Editor
1894   Paul Green. Dramatist
1907   Bobby Jones. Golfing great
1914   Sammy Baugh of football
1918   Mercedes McCambridge. Actress
1919   Nat "King" Cole
1926   Edward Allen. Fifth bishop of Mobile
1936   Thomas Mattingly. Astronaut
1944   John Sebastian
1949   Patrick Duffy. Actor
1952   Susie Allanson. Country singer
1954   Lesley-Anne Down
1959   Danny Ainge of Boston Celtics

March 18

1545   Julius Echter
1585   François Bourgoin
1604   John IV. King of Portugal
1690   Christian Goldback. Mathematician
1702   John Fillmore. Author
1739   Isaac Van Nuys
1745   Elisha Lord. Patriot
1748   John Whiteside
1752   Solomon Tisdale of the Continental Army

1782	John Calhoun. Vice President of the United States
1813	Friedrich Hebbel
1837	Grover Cleveland. President of the United States
1844	Nikolai Rimski-Korsakov. Composer
1852	Rose Coghlan. Actress
1887	Edward Everett Horton
1891	Margaret Banning. Novelist
1891	Joseph Alders. Army chaplain; first bishop of Lansing
1897	Martin Durkin. Secretary of Labor
1901	Mildred Jordan. Author
1914	Judith Arlen. Actress
1929	Andy Granatelli. Racing great
1931	Martha Hickey
1932	John Updike. Author
1939	Charley Pride. Singer
1943	Kevin Dobson. Actor
1956	Ingemar Stenmark. World Cup Winner
1963	Vanessa Williams. Singer

March 19

1534	José de Anchieta. Missionary; "The Apostle of Brazil"
1601	Alonso Cano. Painter of the Spanish baroque period
1720	Gen. George Godfrey
1751	Joseph Lindsmith. Patriot
1768	François Basio. Sculptor
1804	Edward Dana
1807	Anthony Rey. American chaplain during Mexican War
1813	David Livingston. Missionary; explorer
1839	Frank Atkinson. Actor
1848	Wyatt Earp. Gunfighter
1850	Alice French. Author
1857	Edward Robb. Congressman
1860	William Jennings Bryan
1864	Charles Marion Russell. Painter
1874	William Bagley. Author
1879	Joseph Haas. Composer
1881	Edith Nourse Rogers. Thirty-five year congresswoman
1888	Joseph Albers. Painter
1892	Gen. James Van Fleet. Commander U.S. Forces
1900	Jean Joliet. Scientist

1900	Bishop Joseph Annabring of Superior
1901	Jo Mulzener. Designer
1907	Elizabeth Maconchy. Composer
1916	Irving Wallace. Writer
1944	Linda Bird Johnson
1950	Tony Adams of San Diego Chargers

March 20

1647	Jean de Houtefemmle. Scientist
1668	Col. Daniel Sherwood
1739	Nathaniel Fillmore of Revolutionary War; grandfather of President Millard Fillmore
1758	Uriah Corning of the Colonies
1770	Friedrich Halderlin. Poet
1774	John Braham. Tenor
1780	Thomas Metcalfe. Congressman; senator
1795	Becca Gage. Early settler
1799	Simon William Brute. Missionary to Potawatomi Nation
1800	Thomas Webster. Painter
1816	Benedict Sestini. Writer
1823	Ned Buntline. Writer
1825	Cornelia Spencer. Historian
1848	Albert Ryder
1868	Joseph Kirlin. Author
1890	Ben Gigli. Singer
1890	Lauritz Melchior. Tenor
1897	Frank Sheed. Publisher
1907	Hugh McLennan. Writer
1909	Kathryn Forbes. Novelist
1931	Hal Linden. Singer
1948	Bobby Orr of ice hockey
1957	Spike Lee. Actor
1957	Theresa Russell. Actress

March 21

1295	Bl. Henry Suso. Writer; mystic; author of *The Little Book of Eternal Wisdom*
1317	Isabel De Verdon of Amitbury

1466	Charitas Pirkheimer. Abbess
1474	St. Angela Merici. Foundress of the Ursilines
1595	Ferdinando Ughelli. Historian
1609	John II. King of Poland
1679	Benedict Leonard. Fourth Lord Baltimore
1685	Johann Sebastian Bach. Composer
1736	Claude Ledaux. Architect
1763	William MacNeven. Writer; editor
1770	Jean Bureau. Schoolmaster; Postmaster General
1790	Jane Hilton
1812	William George Ward
1837	Theodore Gill. Scientist
1838	Gen. Martin McMahon
1844	Charles Widor. Composer
1861	Alber Chavalier. Singer
1867	Charles Baldwin. Author
1867	Florenz Ziegfeld of the Follies
1878	Pasquale Amato. Baritone
1881	Raymond Hood. Architect
1882	Gilbert Anderson. Actor
1884	Francis Walsh. Editor; author
1905	Phyllis McGinley. Poet
1926	Richard Longshore. Legislator

March 22

1365	Sir Thomas Mowbray. Earl of Nottingham
1459	Emperor Maximilian I. Holy Roman Emperor
1599	Anthony von Dyck. Flemish painter
1762	Hezekia Baratt. Patriot
1785	Adam Sedgwick. Geologist
1788	Pierre Pelletier. Scientist
1813	Thomas Crawford. Sculptor
1815	August Gemunder of violin fame
1816	John Kensett. Painter
1817	Braxton Bragg. Confederate general
1819	William Elder. Third bishop of Natchez
1845	John Bannister Tabb. Poet
1859	Sir Arthur Conan Doyle. Author of *Sherlock Holmes*
1862	Laura Libbey. Author
1868	Robert Millikan. American physicist

1884	Arthur Vandenburg
1891	Chico Marx
1899	Ruth Page. Dancer
1903	Ellin Berlin. Novelist
1907	Stuart Reynolds. Producer
1907	Rober Bhin. Director
1908	Louis L'Amour. Author
1909	Gabrielle Roy. Novelist
1912	Agnes Martin. Painter
1912	Wilfrid Brambell. Actor
1914	Karl Malden. Actor
1920	Werner Klemperor. Actor
1922	Marcel Marceau. Performer
1931	Burton Richter. Recipient of Nobel Prize
1964	Melody Sue. Musician

March 23

1429	Margaret of Anjou
1699	John Bartram. Botanist
1709	Zipporah Kinney of the Colonies
1730	Thomas Love. Patriot
1736	Arthur St. Clair. First governor of Northwest Territory
1772	David Stone
1812	Stephen Return Riggs
1823	Schuyler Colfax. Vice President of the United States
1826	Ludwig Minkus. Composer
1854	Alfred Milner. Statesman
1857	Fannie Farmer of Boston Cooking School
1861	Francis Bourne
1881	Roger Martin du Gard. Writer
1882	Emmy Noether. Mathematician
1898	Robert Ames. Actor
1899	Louis Adamic. Novelist
1903	Daniel Brick. Editor
1907	Daniele Bouet. Nobel Prize winner
1908	Joan Crawford
1910	Akira Kurosowa of "The Seven Samurai"
1912	Wernher von Braun. Rocket expert
1921	Donald Campbell. Auto racer
1927	Regina Crespin. Soprano

1929	Eugenia McCrary. Writer
1933	Norman Bailey. Baritone
1937	Craig Breedlove. Race car driver
1954	Moses Malone of basketball
1990	Princess Eugenie of York

March 24

1278	Lucy de Thwenge of England
1335	Sir Edward Despenser
1494	Georgius Agricola. Mining expert
1630	Joseph Aquirre
1693	John Harrison. Inventor
1703	José Isla
1754	Joel Barlow. Poet
1755	Rufus King. A framer of the Constitution
1797	Antonio Rosmini. Founder of the Institute of Charity
1801	Moses Gregory. Legislator
1813	Taylor Buffington
1816	Andrew Loop
1821	Elisa Felix. Actress
1842	Gabrielle Krauss. Soprano
1851	Garrett Serviss. Writer
1862	Frank Benson. Painter
1872	Eugene Moran. Writer
1874	Joseph Skelly
1876	Frank O'Hara. Editor
1893	Walter Baade. Astronomer
1897	Lucia Chase. Ballerina
1902	Thomas Dewey
1903	Malcolm Muggeridge. Author
1906	Bob Adler. Actor
1909	Clyde Barrow of Bonnie and Clyde
1917	John Kendrew. Scientist
1924	Lois Andress
1924	Kenneth O'Donnell
1951	Pat Bradley of golf

March 25

| 1275 | Sir Henry Percy. Baron of Alnwick |
| 1347 | St. Catherine of Siene |

1434	Bl. Eustochia Calafato
1593	St. Jean de Brehenf. Missionary to the Huron
1598	Ven. Ralph Corbington. Martyr
1633	Samuel Whiting of Lincolnshire
1640	John Stebbins
1648	Basilio Brollo
1699	Johann Hasse. Composer
1752	Zachariah Blankenheckler
1769	Salvator Vigano. Choreographer
1783	Luther Rice
1797	James Ramsey. Author
1818	Elizabeth Ann Umstead
1821	Isabella Banks
1827	Adm. Stephen Luce. A founder of the Naval War College
1837	Patrick O'Rorke. Hero of the Civil War
1862	George Sutherland. Supreme Court Justice
1867	Arturo Toscanini. Conductor
1871	Gritzon Borglum. Sculptor of Mt. Rushmore
1872	Toson Shimazaki. Poet
1881	Bela Bartak. Composer; pianist
1901	Raymond Firta. Anthropologist
1914	Norman Borlaug. Winner of Nobel Prize
1920	Howard Cosell
1925	Flannery O'Connor. Author
1928	James Lovell. Astronaut
1938	Hoyt Axton. Singer
1942	Aretha Franklin
1946	Bonnie Bedelia. Actress
1947	Elton John

March 26

1414	Thomas De Clifford. 8th Lord Clifford
1478	Hieronymus Emser. Writer
1673	Joseph Cassani. Chronicler
1699	Hubert Gravelot. Illustrator
1741	Jean Moreau. Painter
1743	Moses Robinson. Governor of Vermont
1749	Nehemiah Butler. Patriot
1749	William Blount. Signer of U.S. Constitution
1773	Nathaniel Bowditch. Astronomer

1780   Moses Stuart
1781   Nathaniel Gould. Musician
1807   Bl. Bogdan Janski
1813   Gen. Thomas West Sherman of the Union Army
1817   Herman Hocipt of railroad fame
1827   Jules Duprato. Composer
1850   Edward Bellamy. Author
1850   Levicy Ann Spratt
1851   Andrew Bradley. Shakespearean critic
1859   Alfred Houseman. Poet
1865   Cornelius Shear. Botanist
1874   Oskar Nedball. Composer
1875   Robert Frost. Poet
1884   Wilhelm Backhous. Pianist
1895   Adm. Robert Carney. Chief of Naval Operations
1904   Joseph Campbell. Writer
1905   André Cluytens. Conductor
1919   Tennessee Williams
1925   Pierre Boulez. Composer
1925   Vesta Roy. Governor of New Hampshire
1925   Claudio Spies. Composer
1926   Betty Zoe Bass. Artist
1930   Justice Sandra Day O'Connor
1931   Leonard Nimoy. Actor
1934   Alan Arkin
1939   Phillip Allen. Actor
1944   Diana Ross
1950   Teddy Pendergrass
1954   Leslie Amper. Pianist
1959   William Aldrich. Songwriter

March 27

1562   Jakob Gretser. Scholar
1666   Gurdon Saltonstall. Governor of Connecticut
1688   Marco Cornelio Bentivoglio
1703   Johann Eberlin. Composer
1714   Francesco Zaccaria. Historian
1724   Jane Colden. Botanist
1746   Seth Crocker. Patriot
1799   Alfred Vigny. Poet

1841	Grove Johnson. Congressman
1843	Ira Dutton
1843	Elijah McCoy. Inventor
1844	Adolphus Greeley. Arctic explorer
1845	Wilhelm Roentgen. X-ray pioneer
1847	Otto Wallach. Pulitzer Prize winner
1855	Vincent d'Indy. Composer
1857	Eustace Shaw. Editor
1863	Sir Henry Royce. Founder of Rolls Royce
1867	Edyth Walker. Opera singer
1869	Adm. Henry Ziegameier
1881	Mariska Aldrich. Opera singer
1886	Ludwig Mies von der Rahe. Founder of modern architecture
1899	Gloria Swanson
1914	Budd Schulberg. Novelist
1923	Louis Simpson. Poet
1924	Sarah Vaughn. Singer
1940	Cale Yarborough. Winner of Daytona, the Atlanta 500 and the Southern 500
1941	Charles Pashayan. Congressman

March 28

1472	Fra Bartolommeo. Renaissance painter
1515	St. Teresa of Avila. Doctor of the Church
1522	Albert. Margrave of Brandenburg
1652	Samuel Sewell of the Salem Witch trials
1660	George I. King of England
1702	Girolamo Ballerini. Scholar
1731	Ramón Cruz. Poet
1749	Pierre Laplace. Astronomer
1793	Henry Schoolcraft. Explorer
1811	St. John Newmann
1818	Wade Hampton of the Civil War
1826	Philip Wolfersberger
1840	Pasha Emin. Explorer
1866	Max Bendix. Orchestra Conductor
1867	Edmond Clement. Tenor
1873	Anne Sedgwick. Novelist
1875	Helen Westley. Actress
1878	Herbert Lehman. Governor of New York

1900 Moses Yellowhorse of Pittsburgh Pirates
1902 Corliss Lamont. Writer
1902 Flora Robson. Actress
1903 Rudolf Serkin. Pianist
1906 Murray Adasken. Composer
1910 Queen Ingrid of Denmark
1920 John Weaver. Sculptor
1922 Grace Hartigan. Painter
1924 Freddie Bartholomew. Actor
1954 Reba McEntire

March 29

1702 Ven. Caesar Sportelli
1724 Job Crocker of the Colonies
1724 Zebulon Brevard. Patriot
1746 Carlo Bonaparte. Father of Napoleon
1773 Franz Hladnik. Scientist
1788 Charles V (Don Carlos) of Spain
1790 James Tyler. President of the United States
1803 William Gowans. Antiquarian
1803 John Milligan
1804 Nehemiah Abbott of Congress
1807 Karoline Bauer. Actress
1823 Signey Frances Bateman. Actress
1831 Amelia Barr. Novelist
1861 Herman Bemberg. Composer
1865 Stephen Bonsal. Newspaperman
1867 Cy Young of baseball
1869 Edwin Lutyens. Composer
1887 Julius Becker. Poet
1889 Howard Lindsay. Playwright
1892 Jozsef Cardinal Mindszenty. Renowned for his uncompromising opposition to fascism, nazism and communion
1906 Power Biggs. Organist
1911 Philip Ahn. Actor
1918 Pearl Bailey. Singer
1931 Gloria Davy. Soprano
1937 Billy Carter

March 30

- 1394 Thomas De Knightley de Charlton of England
- 1680 Angelo Quirini
- 1734 Thomas McKean. Signer of the Declaration of Independence
- 1753 Nathaniel Cushing. Patriot
- 1793 Juan Manuel Rosas
- 1798 Luise Hensel. Early settler
- 1820 Francis Baker
- 1820 James Corcoran. Editor
- 1832 Roger Mills. Confederate Officer wounded at Mission Ridge; representative
- 1842 Thomas Middleton. Historian; writer; editor
- 1842 John Fiske. Historian
- 1844 Paul Verlaine. Poet
- 1848 Albinus Nance. Governor of Nebraska
- 1848 Carlos. Duke of Madrid
- 1849 Howard W. Smith
- 1853 Vincent van Gogh
- 1854 Willard Vandiver. Congressman
- 1858 De Wolf Hopper
- 1862 Umberto Benigni
- 1883 Jo Davidson. Artist
- 1906 John Randall. Editor
- 1922 Francis Akos. Violinist
- 1930 John Astin. Actor
- 1938 Warren Beatty
- 1953 Steve Peace. Legislator
- 1969 Nina Khalil. Singer; dancer; businesswoman

March 31

- 1499 Pope Pius IV (Giovanni Angelo de' Medici)
- 1621 Andrew Marvell. Poet
- 1675 Pope Benedict XIV (Prospero Lorenzo Lambertini)
- 1684 Francesco Durante. Composer
- 1709 Capt. Israel Nickerson
- 1732 Franz Joseph Haydn. Composer
- 1757 Cornelius Linderman. Patriot
- 1762 John Weaver of the Colonies
- 1804 Dean Richmond. Railroad tycoon whose corporation merged into the New York Central

1806	John Parker Hale. Senator
1811	Robert Bunson. Invented bunson burner
1813	Auguste Careyon. Author
1824	William Hunt. Painter
1835	John La Farge. Painter
1844	Andrew Lang. Writer
1851	John Calhoun Tutt. Writer
1855	Gilbert Gaul. Artist
1857	Pope Pius XI (Achille Ratti)
1878	Jack Johnson. Heavyweight
1880	Sean O'Casey. Author
1889	Muriel Hazel Wright. Historian
1895	Vardis Albero Fisher. Writer
1900	Prince Henry. Duke of Gloucester
1907	Eddie Quillan. Actress
1912	Iva Kitchell. Dancer
1923	Ellsworth Kelly. Painter
1928	Medardo Rosso. Sculptor
1928	Gordie Howe of hockey
1928	Lefty Frizzell
1929	Liz Claiborne

April 1

1282	Louis IV. King of Germany; Holy Roman Emperor
1588	Ven. Robert Drury
1613	Guilo Bartolocci. Writer
1621	Giuseppe Agnelli
1688	Maurus Dantino
1697	Antoine Provost. Novelist
1746	Michael Levadoux. Missionary
1755	Samuel Fleming. Patriot
1762	Seth Ligon of the Revolutionary War
1801	William Lynch. Explorer
1808	Nelson Barrere. Legislator
1820	James McMaster. Editor
1823	Simon Bolivar Buckner. Confederate general; governor of Kentucky
1835	James Fisk, Jr. of Erie Railroad
1836	Lucretia Winans
1852	Edwin Abbey. Painter

1855	Agnes Repplier. Author
1866	Ferruccio Busoni
1868	Edmond Rostand. Poet
1873	Sergei Rachmaninoff. Composer; piano virtuoso
1875	Edgar Wallace. Novelist
1882	Florence Blanchfield. Superintendent of Army Nurse Corps
1883	Lon Chaney
1886	Wallace Beery. Actor
1894	Albert Beck. Composer
1931	George Baker. Actor
1932	Debbie Reynolds
1935	William Turnbull. Architect
1935	Charles Ainsworth. Writer
1965	Mark Jackson of basketball

April 2

742	Charlemagne
1315	Simon deGhent. Scholar
1586	Pietro Della Valle
1602	Ven. Mary of Agreda
1613	Francesco Grimaldi. Astronomer
1617	Anna Sibylla. Painter
1645	François Belmont. Missionary
1745	Francis Triesnecker. Astronomer
1745	Richard Bassett. Revolutionary statesman
1752	Silas Bullard. Patriot
1769	Elizabeth Bouder. Served in Revolutionary War
1805	Hans Christian Anderson
1806	Giacomo Antonelli
1816	George Hay Stuart of YMCA
1817	Erastus Palmer. Sculptor
1834	Frédéric Auguste Bartholdi. Created Statue of Liberty
1861	Ernest van Dyck. Tenor
1862	William B. Wilson. First U.S. Secretary of Labor
1865	Irving Babbitt
1877	Mordecai Ham. Radio preacher
1906	Adm. William S. Grant, Jr. Pacific Commander during Vietnam War
1908	Buddy Ebsen
1911	Jeremy Ingalls. Poet

1914 Sir Alec Guinness
1920 Jack Webb

April 3

1110 Thiofrid of Echternach. Reformer; author
1367 Henry IV. King of England
1744 Richard Taylor of Revolutionary War; Collector of Internal Revenue; father of President Zachary Taylor
1783 Washington Irving
1798 John Banim. Novelist
1798 Charles Wilkes. Explorer
1822 Edward Everett Hale. Writer; chaplain of U.S. Senate; wrote *The Man Without a Country*
1823 William Tweed. Politician
1826 Cyrus K. Halliday. Legislator; major figure in the history of the Atchison, Topeka and Santa Fe Railway
1863 Henry van de Velde. Painter; architect
1865 Herbert Vivian. Writer
1870 Sarah Conboy. Labor leader
1876 Margaret Mary Anglin. Actress
1879 Augusta Seaman. Writer
1888 Adm. Thomas Kincaid. Commander, Seventh Fleet; directed Aleutians campaign
1893 Leslie Howard. Actor
1898 George Jessel. Actor
1923 Jan Sterling. Actress
1924 Doris Day
1926 Alfred Thompson. Winner of National 400
1926 Virgil Grissom. Astronaut
1928 Kerstin Meyer. Mezzo soprano
1930 Helmut Kohl
1933 Robert Dornan of Congress
1941 Stanley Wanlass. Sculptor; painter
1955 James Smith. Heavyweight
1961 Eddy Murphy

April 4

188   Marcus Aurelius. Roman Emperor
1527 Abraham Ortelus. Publisher of first atlas

1588	Thomas Hobbs
1648	Grinling Gibbons
1718	Benjamin Kennecott of the Colonies
1734	Stephen Sewall. Representative; publisher
1748	William White. Patriot
1752	Niccolo Zingarelli. Composer
1762	Solomon Livermore
1780	Edward Hicks. Painter
1786	Cheltan Allan. Congressman
1788	David Gouverneur Burnet. First president of the Republic of Texas
1813	James Aylward. Poet
1819	Maria II. Queen of Portugal
1821	Linus Yale. Inventor of Yale lock
1838	Laurence Barretto. Actor
1844	John Riley Tanner. Pioneer
1875	Pierre Monteux. Conductor
1876	Maurice de Vlaminck. Painter
1886	Victor Ridder. Publisher
1888	Tris Speaker. Played for four teams during a 22-year major league career
1889	Todd Wehr
1891	Butts Butler of National Football Hall of Fame
1896	Robert Sherwood. Playwright
1899	Amy Jones. Artist
1906	Janet Gaynor
1908	Edward Murrow
1908	Antony Tudor
1913	Jerome Weidman
1914	Lucy Crockett. Author
1915	Lars Ahlin. Novelist
1915	Muddy Waters. Blues singer
1916	Giorgio Bassani
1932	Anthony Perkins. Actor
1938	Angelo Giametti. Baseball Commissioner

April 5

1516	Walter Strickland of Sizergh
1568	Pope Urban VIII
1622	Vincenzo Viviani. Mathematician
1626	Frances Chichester of Braunton

1698	George Wagner. Violinist; composer
1710	Marie Carmago. Ballerina
1726	Benjamin Harrison. Signer of the Declaration of Independence. Father of President William Henry Harrison and great-grandfather of President Benjamin Harrison
1761	Sybil Ludington. Heroine of Revolutionary War
1784	Ludwig Spoker. Violinist; conductor; composer
1809	John van Nordwick of Chicago, Burlington & Quincy Railroad
1809	George Selwyn
1816	Samuel Miller. Supreme Court Justice
1827	Joseph Lister. Antiseptic pioneer
1834	Frances Stockton. Writer
1835	Adm. Francis Ramsey. Commander of New York Navy Yard
1836	John Raymond. Actor
1853	Henry St. George Tucker of Congress
1855	William Carey Jones. Free silver advocate
1856	Booker T. Washington of Tuskegee Institute
1869	Albert Roussel. Composer
1871	Pop Warner of football
1874	Jesse Holman Jones. Secretary of Commerce
1893	Alessio De Paolis. Tenor
1900	Spencer Tracy
1908	Bette Davis
1908	Herbert von Karajan. Conductor
1916	Gregory Peck
1920	Arthur Haley
1922	Gale Storm

April 6

1483	Raphael, the great painter of the High Renaissance
1651	André Dacier
1651	Joseph Whiting of Southhampton
1745	William Dawes. Rode with Paul Revere
1755	George Longenberger
1756	Samuel Brandenburg
1809	Alfred Lord Tennyson
1809	Prince Frederick Schwarzenberg
1810	Philip Gosse. Zoologist
1818	A.O. Vinje. Poet
1821	Sparrow Howes. Pioneer

1838	Kate Wells. Writer
1850	Nicholas Matz. Bishop of Denver
1862	Fitz-James O'Brien. Playwright
1867	Butch Cassidy
1869	William Hale. Writer
1875	General Henry Butner
1885	Parry Thomas. Holder of Land Speed Record
1890	Millard Tydings. Senator
1892	Donald Douglas. Developer of DC-3
1909	Herman Long of racing
1923	Herb Thomas. Championship race car driver
1926	Patricia Aldrich. Editor
1929	André Previn. Conductor
1937	Merle Haggard

April 7

1506	St. Francis Xavier. Missionary to the Orient
1640	Louis Hennepin. Explorer
1652	Pope Clement XII (Lorenzo Corsini)
1735	Pierre Bibault. Missionary
1755	Jedediah Lee. Patriot
1770	William Wordsworth. Writer
1773	John Wayles Eppes. Congressman; senator
1775	Lucy Grennell. Pioneer
1777	Orange Mervin. Senator
1786	William King. Vice President of the United States
1807	Luther Cullender
1825	John Henry Gear. Governor of Iowa
1847	William Pitt Bradshaw. Statesman
1848	Alexander Tyler of Franco-Prussian War; son of President Tyler
1860	W.K. Kellogg. Creater of Corn Flakes
1873	John McGraw of Baltimore Orioles
1881	James MacDonnell. Poet
1884	Homer Klein
1889	Gabriela Mistral. Poet
1890	Marjory Stoneman Douglas. Author
1890	Harry Hill. Amphibious leader of World War II; governor Naval Home
1893	Irene Castle. Dancer
1896	Sherman Fairchild. Inventor

1908	Frank Fitzsimmons
1915	Billie Holiday. Jazz singer
1924	Ikuma Dan. Composer
1928	James Garner. Actor
1933	Wayne Rogers
1935	Bobby Bare
1936	Preston Jones
1939	Frank Coppola. Director

April 8

1338	Stephen of Gravesend
1692	Guiseppi Tortini. Violinist; composer
1726	Lewis Morris. Signer of Declaration of Independence; delegate to Continental Congress
1732	David Ritterhouse. Astronomer
1771	William Rabun. Governor of Georgia
1763	Benjamin Lufkin of the Colonies
1790	Fitz-Greene Halleck. Author
1793	Karl Zoll. Statesman
1811	Theodore Woodruff. Inventor; wagon builder; railroad car builder
1832	Howell Edmunds Jackson. Senator; Supreme Court Justice
1869	Mabel Louise Fuller. Author
1875	Albert I. King of the Belgians
1886	Margaret Barnes. Novelist
1888	Frank Shay. Publisher; writer
1889	Sir Adrian Boult
1902	Joseph Krips. Conductor
1905	Ilka Chase
1905	Reg Manning. Cartoonist
1906	Raoul Jabin. Tenor
1911	Paul Hallinan. First archbishop of Atlanta
1912	Sonja Henie
1918	Betty Ford
1919	Ian Smith. Statesman
1923	Franco Corelli. Tenor
1926	Sue Casey. Actress
1929	Jaques Brel. Songwriter
1940	John Havlicik of Boston Celtics

April 9

1458	Bl. Baptista Veirani. Writer
1483	Paulus Jovius. Historian
1662	Edward Hawarden
1737	Joshua Ricker of the Colonies
1758	Fisher Ames. Secured passage of Jay Treaty
1759	Adrian Davenport. Patriot
1821	Charles Pierre Baudelaire. Poet
1823	Lorenzo Coffin. Land agent; grange agent; Iowa Railroad Commissioner
1827	Maria Cummins. Author
1831	Geremia Bonomelli. Writer
1835	Leopold II. King of the Belgians
1838	Henry Richter. First bishop of Grand Rapids
1842	Henry Tuck
1846	Francesco Paola Tasti. Composer
1859	Edward Burt. Botanist
1862	Charles Henry Brent
1864	Frank Tubbs. Author
1893	Michael Ready
1903	Marjorie Rhodes. Actress
1905	Senator J. William Fulbright
1908	Victor Vasarely. Painter
1909	Robert Helpmann. Actor
1909	Ivan Dzerzkinsky. Composer
1943	Jeanne Pierre Bonnefoux of ballet
1954	Dennis Quaid. Actor

April 10

401	Theodosius II. Roman Emperor in the East
1460	Bl. Anthony Neyrot
1512	James V. King of Scotland
1696	Esther Wheelwright
1752	Alexander Graydon. Author
1767	William Warren. Actor
1794	Commodore Matthew Perry
1805	James Fitton. Missionary; writer
1806	Leonidas Polk. Confederate general; killed at Pine Mountain
1810	Benjamin Day. Founder of *New York Sun*

1814	Catharine Wasson Hall
1827	Lew Wallace. General; served in Mexican War, West Virginia campaign and Shiloh; governor of New Mexico; author
1829	William Booth. Founder of Salvation Army
1830	Cyrus Wilson Riley
1835	Henry Villard. Journalist; correspondent
1838	Frank Baldwin. Inventor
1847	Joseph Pulitzer. Established Pulitzer Prize
1848	John Kenna. Senator
1880	Frances Perkins. Secretary of Labor
1891	Tim McCoy. Actor
1898	Horace Gregory. Poet
1903	Clare Booth Luce
1921	Sheb Wooley. Singer
1924	Chuck Conners. Actor
1929	Mike Hawthorn. Winner of World Drivers' Championship
1929	Max Von Sydow. Actor
1932	Omar Sharif. Actor
1938	Glen Campbell. Singer

April 11

1357	John I. King of Portugal
1492	Margaret of Navarre
1586	Pietro Valle. World traveler; author
1698	Lydia Day
1721	David Zeisberger. Missionary
1744	John Fiske. Captain of the "Tyronnicide" during Revolutionary War
1749	Adelaide Labille-Guiard. Painter
1757	Aaron Putnam
1758	William Loving
1767	Jean Baptiste Isabey. Artist
1774	George Haydock. Scholar
1800	Joseph Deharbe. Author
1810	Johan Kulschker
1811	Oliver Rouge. Explorer
1818	Thomas Doe of War of 1812
1836	William DuBose. Chaplain of Confederate Army; author
1839	General St. Clair Mulholland. Recipient of Congressional Medal of Honor

1862	Charles Evans Hughes. Chief Justice of Supreme Court
1863	Henry Balfaur. Anthropologist
1884	Genevieve Macaulay
1893	John Nash. Painter
1900	Rosalind Atkinson. Actress
1901	Donald Menzel. Astronomer
1902	Muriel Elwood. Writer
1902	Quentin Reynolds. Journalist
1904	Paul McGrath. Actor
1913	Oleg Cassini
1914	Norman McLaren
1919	Hugh Carey. Governor of New York

April 12

1573	Jaques Bonfrere. Scholar
1578	Laurent Beyerlinch. Historian
1692	Giuseppe Tartini. Composer
1722	Pietro Nartini. Musician
1759	Ebenezer Crosby. Patriot
1777	Henry Clay. Senator; Secretary of State
1787	Joseph Ritter. Historian
1795	Charles Morgan. Leader in ship and rail transportation
1789	St. Anthony Gianelli
1801	Capt. George Washington Adams. Legislator; son of President John Quincy Adams; grandson of President John Adams
1802	Ven. Francis Liberman
1808	Pauline Craven. Writer
1818	Michael Heiss
1831	Grenville Dodge. Railroader, banker; engineer
1831	Constantin Meunier. Painter
1840	Franz Haberl. Historian; editor
1844	Mollie Davis. Author
1848	James Q.W. Wilhite
1851	Katharine Bates. Poet
1857	John Thomas Underwood. Early manufacturer of typewriters
1865	James Joseph Walsh. Author; edited *A Golden Treasury of Medieval Culture*
1880	Wayne Arey. Actor
1899	Gladys Taber. Author
1901	Joseph Plumpe. Editor

1904	Lily Pons. Soprano
1913	Lionel Hampton
1924	Curtis Turner. Winner of Southern 500 and Permatex 300
1932	Jack Gelber. Playwright
1933	Montserrat Caball. Soprano
1942	Carlos Reutemann. Race car driver
1943	Charles Ludlam. Actor
1947	Emmylou Harris
1950	David Cassidy

April 13

1506	Bl. Peter Faber
1519	Catherine de Medici
1570	Guy Fawkes
1628	Peter Lambeck. Historian
1648	Jeanne Guyon. Writer
1652	Thomas Ward. Poet
1699	Alexander Ross. Poet
1710	Jonathan Carver. Explorer of Great Lakes region
1721	John Hanson. Revolutionary War leader
1739	Bezalel Loring. Patriot
1740	Samuel Bradford
1743	Thomas Jefferson. Signer of Declaration of Independence and President of the United States
1748	Joseph Bramah. Inventor of hydraulic press
1750	John Trumbull. Poet
1769	Thomas Laurence. Artist
1771	Richard Trevithick. Inventor
1794	John Kenedy. Publisher
1810	Félicien David. Composer
1825	Thomas McGee. Editor
1835	Louis Lambert
1842	Courtney Warren
1845	Herman Alerding. Bishop of Ft. Wayne; historian
1854	William Drummond. Poet
1873	John Davis. Presidential candidate
1873	Edie Hall
1875	James Jeffries. Heavyweight champion
1883	Effigene Wingo. Congresswoman
1890	Frank Murphy. Governor of Michigan; Supreme Court Justice

1892 Robert Watt. Radar pioneer
1905 Bruno Rossi. Scientist
1911 Pietro di Donato
1919 Howard Keel. Actor
1926 Don Adams. Actor
1931 Dan Giurney. Race car driver; winner of La Mans and Grand Prix
1963 Gary Kasparov. World Champion Chess Player

April 14

1547 Johannes De Buyus. Author
1741 Hezekia Root. Patriot
1802 Horace Bushnell
1807 Simon Davis
1810 Lot Myrick Morrell. Secretary of Treasury; governor
1824 Mandana Thomson
1827 Augustus Pitt-Rivers. Archeologist
1840 Richard Burtsell
1857 Edgar Kelley. Composer
1857 Princess Beatrice Marie Victoria of Battenberg
1864 Allan Aynesworth. Actor
1879 James Branch Caball. Novelist
1892 Juan Belmonte. Bullfighter
1900 Salvatore Baccaloni. Basso
1907 François Duvalier "Papa Doc." President of Haiti
1916 Emerson Buckley. Conductor
1925 Jerome Reppa. Legislator
1925 Rod Steiger
1925 Joseph Mattioli of racing
1932 Anthony Perkins
1940 Julie Christie
1941 Connie Smith. Singer
1941 Pete Rose
1942 Valeri Brumel. High jumper
1949 John Shea. Actor

April 15

1340 Angelo Cardinal Acciaioli. Archbishop of Florence
1452 Leonardo da Vinci. Artist

1707	Leonhard Euler. Scientist
1726	Jonathan Walker of the Colonies
1741	Charles Wilson Peale. Portrait painter; founder of Peale's Museum
1748	Stephen Longwell. Patriot
1757	Caleb Bingham of the Revolution
1762	Cornelius Darnell of the Colonies
1782	Eleazar Wheelock Ripley of Congress
1784	Fanny Gage. Pioneer
1800	James Ross. Discovered North Magnetic Pole
1812	Theodore Rousseau. Painter
1814	John Lathrop Motley
1823	Lucy Skinner
1841	Henry Aplin. Congressman
1843	Henry James
1844	George Glidden of the Navy
1859	George Burr. Artist
1867	Dooley Brainard
1889	A. Philip Randolph. Labor leader
1891	Georges Roesch. Leading automobile designer
1894	Bessie Smith. Blues singer
1895	Corrado Alvaro. Novelist
1901	Joe Davis. World Snooker Champion
1933	Ray Clark. Singer
1933	Mel Kenyon

April 16

1220	Ambrose of Siena
1319	John II. King of France
1612	Abraham Calav
1661	Charles Montague. Statesman
1769	Peter Brakehill of the Colonies
1779	Giovanni Inghirami. Astronomer
1783	Ven. Joaquina de Mas
1804	Francis McFarland. Bishop of Hartford
1821	Ford Brown. Painter
1844	Anatole France. Writer
1850	Sidney Thomas. Inventor
1865	Grace Hill. Novelist
1889	Charlie Chaplin
1890	André Danjou. Astronomer

1894	Jerzy Neymar. Mathematician
1897	Milton Cross of radio
1921	Peter Ustinov
1924	Henry Mancini
1925	Marion Montgomery. Writer
1927	Joseph Cardinal Ratzinger
1931	Edie Adams. Actress
1935	Bobby Vinton
1935	Jigger Sirois. Race driver
1940	Queen Margaret of Denmark
1947	Kareem Abdul-Jabbar
1949	Richard Cowden-Guido. Writer

April 17

1575	Maximilian. Duke of Bavaria
1578	Max Van der Sandt
1620	Bl. Margarite Bourgeoys
1622	Henry Vaughan. Poet
1741	Samuel Chase. Signer of the Declaration of Independence
1741	Johann Gottlieb Naumann. Composer
1744	Job Cushing. Patriot
1745	Dudley Currier
1772	Archibald Alexander. President of Hampton-Sidney College
1786	Davy Crockett
1797	John Ogilvie
1813	Susan Cooper. Author
1818	Emperor Alexander II
1820	Alexander Cartwright. "Father of Modern Baseball"
1821	Ambrose Ranney of Congress
1824	William Spohn Baker. Author
1835	John Pierpont Morgan
1845	Katherine Barrows. Editor
1849	William Rufus Day
1859	Walter Camp. "Father of American Football"; founded Intercollegiate Football Association
1859	Willis Van Devanter. Supreme Court Justice
1868	Adm. Mark Bristol. Commander-in-chief Adriatic Fleet
1874	Clarence Mackay. Laid transatlantic cable
1882	Artur Schnabel. Pianist; composer
1889	Claude Abbott. Writer

1897 Thornton Wilder. Writer
1918 William Holden. Actor

April 18

1161 Theobold. Archbishop of Canterbury
1480 Lucrezia Borgia
1580 Thomas Middleton. Writer
1641 Agatha Lauritzen
1721 Shearashub Bourne
1728 Ebenezer Learned. Revolutionary War hero
1736 Jacob Weidman. Patriot
1753 Thaddeus Birdseye
1762 Moses Lord of the Colonies
1764 Abner Loveland. Patriot
1790 Anton Donnersberger. Landscapist
1816 Mother Euphenia. Superior of the Sisters of Charity in the United States
1819 Franz von Suppé. Composer
1864 Richard Harding Davis
1870 William Orcutt. Author
1882 Leopold Stokowski
1896 Randolph Bandas. Author; columnist
1914 Claire Martin. Writer
1918 Elisabeth Borgese. Author
1939 Jorge Anders. Composer
1939 Marcia Haydee. Ballerina
1942 Franklin Ashley. Writer
1947 James Wood. Actor
1948 Steven Anderson. Writer
1953 Ralph Baggiari. Musician

April 19

1483 Paulus Jovius. Historian
1654 Samuel Sturtevant of Plymouth
1665 Jacques Lelong. Bibliographer
1721 Roger Sherman. Signer of the Declaration of Independence; congressman; senator
1736 Benjamin Huntington of Continental Congress

1761 Philemon Houghton of the Battle of Bennington
1763 Benjamin Dorsett of the Colonies
1805 Charles Coussemaker. Historian
1818 Peter Blenkinsop
1813 David Reid. Senator
1830 Malvina Davis. Actress
1830 Karl Cornely. Scholar
1831 Mary Louise Booth. Author
1832 Lucretia Garfield. First Lady
1858 May Robson. Actress
1864 Prince Pierre Troubetzkoy. Painter
1870 Harry Rousseau. Chief of Bureau of Yards and Docks
1871 Melville Post. Novelist
1877 Gertrude Whitney. Sculptor
1900 George O'Brien. Actor
1905 Adm. Jack Thack. World War II fighter pilot; Air Commander-in-chief U.S. Naval Forces in Europe
1909 Bucky Walters of Boston Braves, Boston Red Sox, Philadelphia Phillies and Cincinnatti Reds
1912 Glenn Seaborg. Scientist
1927 Malcolm Lucas. State Supreme Court Justice
1931 Etheridge Knight. Poet
1932 Dorothy Abbott. Sculptor
1935 Dudley Moore. Actor

April 20

1586 Ven. Richard Sergeant. English martyr
1718 David Brainard. Missionary
1749 Elijah Lovell of the Colonies
1764 Jacob Radcliffe. Mayor of New York; State Superior Court Justice; a founder of Jersey
1783 Alois Buchner
1788 Samuel Baldwin. Settler
1801 Solomon Smith. Comedian
1842 John Cardinal Farley. Archbishop of New York
1843 Rev. Wesley Bates. Served at Cold Harbor, Petersburg and Shenandoah
1850 David French. Sculptor
1858 Lulu Irene Davis
1859 Herbert Squires

1860	Edward C. O'Brien. U.S. Commissioner of Navigation
1861	Amanda Stiles
1866	George Sauter. Painter
1893	Robert Gannon. Author
1893	Harold Lloyd. Comedian
1893	Joan Miro. Artist
1899	Richard McNulty. Administrator of Merchant Marine Cadet Corps
1909	Robert Tallant. Author
1920	James McCargar. Writer
1923	Mother Angelica
1924	Nina Foch. Actress
1930	Harry Agannis of baseball
1941	Ryan O'Neal. Actor
1949	Jessica Lang
1961	Don Mattingly of New York Yankees

April 21

1651	Joseph Vaz. Apostle of Ceylon
1699	Joseph Dow of the Colonies
1729	Catherine the Great
1746	George Van Nostrand
1810	George Phinneas Gordon. Inventor
1816	Charlotte Brontë. Author of *Jane Eyre*
1817	Victor Buck. Scholar
1829	Walter Gilbert. Inventor
1838	John Muir
1846	Alfred Trude. Noted criminal lawyer
1851	James Barron. Commander of the "Chesapeake"
1857	Paul Dresser. Songwriter
1859	Estelle McKenzie
1868	Alfred Maurer. Painter
1870	Edward Ames. Editor of *The Scroll*
1871	Leo Blech. Pianist
1887	Joe "Marse Joe" McCarthy. Manager of New York Yankees, Boston Red Sox and Chicago Cubs; Elected to Baseball Hall of Fame
1893	Richard Skinner. Editor; critic
1899	Randall Thompson. Explorer
1904	Jean Helion. Painter
1911	Leonard Warran. Baritone of the Metropolitan

1915 Anthony Quinn
1920 Burno Moderna. Composer: conductor
1926 Queen Elizabeth II

April 22

1451 Isabella, Queen of Castille; sponsor of Christopher Columbus
1610 Pope Alexander VIII (Pietro Ottoboni)
1658 Giuseppi Torelli. Author of first published violin concerts
1688 Jonathan Dickinson
1707 Henry Fielding. Artist
1711 Eleazar Wheelock of the Colonies
1751 Ephraim Dimond
1775 George Hermes
1803 George Belcourt, Missionary
1839 Olive Logan. Actress
1839 Pauline Weiller. Pianist
1852 Grand Duke William IV of Luxemberg
1857 Ada Rehan. Actress
1874 Ellen Glasgow. Novelist
1876 O.E. Rolvaag. Writer
1891 Vittorio Jano. Race car designer
1908 Eddie Albert
1912 Kathleen Ferrier. Contralto
1916 Yehudi Menuhin. Violinist
1916 Edward Willock. Editor
1922 Charles Mingus. Jazz musician
1923 Barbara Vail. Painter
1923 Gil Ferguson. Legislator
1925 George Cole
1934 Kenneth Maddy. Legislator; state senator

April 23

1371 Sir John Tucket. Lord Audley
1482 Julius Scaliger. Poet
1522 St. Catherine de Ricce
1571 Leon of Modena. Poet
1598 Maarten Tromp of Anglo-Dutch Wars
1756 Nathaniel Luff. Patriot

1775	Joseph Turner. Painter
1791	James Buchanan. President of the United States
1804	Marie Taglioni. Ballerina
1811	John Rice. Governor of Michigan
1812	Louis Jullien. Conductor
1813	Stephen A. Douglas
1818	John Gill Shorter. Governor of Alabama
1834	Chauncey Depew. Senator
1846	Herman Napoleon Coffinbury
1852	Edwin Markham. Poet
1853	Thomas Nelson Page. Historian
1856	Arthur Hadley. Writer
1857	Andrew Rowan. Writer
1864	Richard Knight Le Blond. Industrial leader
1881	Charles Norris. Painter
1888	Daniel Lord. Author; editor
1896	Margaret Kennedy. Author
1899	Edith March. Actress
1921	Janet Blair. Actress
1921	Warren Spahn of Baseball's Hall of Fame
1917	Bob Crosby. Orchestra leader
1928	Shirley Temple
1929	Vera Miles
1932	James Fixx
1934	Barbara Eden
1934	Sonny Jurgensen
1936	Roy Orbison
1942	Sandra Dee
1949	Joyce De Witt. Actress
1949	Shelley Long

April 24

1576	St. Vincent de Paul
1638	Cpl. Samuel Stearns of Watertown
1706	Padre Martini. Composer
1709	George Hadley. Settler
1710	Louis Esglis. Pioneer
1733	David Cutter of the Colonies
1750	John Trumbull. Writer
1783	John Floyd. Congressman; governor of Virginia

1808	Hendrick Wright. Congressman
1815	Anthony Trollope. Novelist
1817	Jean Marignac. Scientist
1825	Robert Ballantyne. Writer
1846	Marcus Clarke
1850	John Lawson Stoddard. Author
1874	John Russell Page
1892	Jack Hulbert. Actor
1903	Mike Michalske. All American
1905	Robert Penn Warren. Novelist
1905	Helen Tamiris. Dancer
1911	Adm. John Coye, Jr. Submarine Commander
1913	Violet Archer. Pianist; composer
1926	Howard Alrecht. Author
1943	Elizabeth Allport. Musician
1954	Robert Carradine

April 25

1214	Louis IX. King of France; leader of Crusades
1284	Edward II. King of England
1284	Sir Roger de Mortimer of England
1683	Markus Hanzik. Historian
1719	Giuseppi Baretti. Writer
1730	Nicolas Baudeau. Economist
1755	Casper Adams. Patriot of the Revolution
1772	Judah Dana. Senator
1811	William Bissell. Governor of Illinois
1817	Marcus Thackaberry. Settler
1839	Thomas Burrill. Botanist
1843	Constance Harrison. Author
1852	Leopoldo Alas. Novelist
1861	Marco Enrico Bossi. Composer
1873	John Chamberlain Ward. Bishop of Leavenworth
1874	Guglielmo Marconi of the wireless
1879	Jenny Heinemann of baking fame
1902	Jimmy Alexander. Actor
1918	Ella Fitzgerald
1918	Astrid Varney. Singer
1919	Douglas Archibald. Novelist
1923	Melissa Hayden

1940 Al Pacino
1946 Talia Shire. Actress

April 26

121 Marcus Aurelius. Roman Emperor
1573 Marie de Medici
1644 Mary Dow
1648 Bathsheba Skiff
1648 Peter II. King of Portugal
1656 John Badman
1718 Esek Hopkins. Commander-in-chief of Continental Navy
1746 Louis Barral. Writer
1747 George Vaughan
1785 John Audubon. Publisher of *The Birds of North America*
1795 Frances Caulkins. Historian
1798 Ferdinand Delacroix
1798 James Beckwourth. Mountaineer; scout
1812 Friedrich von Flowtow. Composer
1813 Mary Hardman. Pioneer
1821 Julius Garesche of the battle of Stone River
1830 Benjamin Franklin Tracy. Secretary of the Navy
1834 Charles Browne. Author
1838 Capt. Robert Bruton of General Custers' Staff
1862 Alexandre Brou. Writer
1879 Owen Richardson. Novel Prize winner
1879 Paul Poiret. Couturier
1886 Gertrude Painey. Blues singer
1893 Anita Loos. Author of *Gentlemen Prefer Blondes*
1898 John Grierson
1900 Charles Richter. Invented the Richter Scale
1902 Jonathan Daniels. Writer
1908 Edward R. Murrow
1912 Alfred Elton Van Vogt. Writer of science fiction
1917 Sal Maglie of baseball
1918 Fanny Blankers-Koen. Olympic track star
1949 Peter Schaufuss of ballet
1950 Lynn McCloud. Writer

April 27

1607 Elizabeth Pelham of Chaldon
1669 Jakob Winslow

1701   Sebastian Redford
1791   Samuel F.B. Morse. Inventor of telegraph
1795   Edward Kavanagh. Governor of Maine
1811   John Oertel. Journalist
1818   Amasa Stone. Railroader
1822   Ulysses S. Grant. President of the United States
1839   Emma Kimball
1840   Rossiter Raymond. Author
1840   Edward Whymper. Explorer
1861   Archelaus Ewing Turner of Chautanqua Association
1865   Charles Dawes. Vice President of the United States
1871   Arthur Nevin. Composer
1874   Maurice Baring. Novelist
1896   Rogers Hornsby of baseball
1898   Ludwig Bemelmans. Painter
1913   Richard Bissell. Playwright
1922   Daphne Anderson. Actress
1922   Jack Klugman
1925   Samuel Baron. Flutist
1938   Earl Anthony of Bowlers Association Hall of Fame
1941   Judith Blegen. Soprano
1943   Helmut Marco. Race driver

April 28

1442   Edward IV. First monarch of the House of York
1738   Armwell Lockwood. Patriot
1757   Austin Corley. Patriot
1758   James Monroe. Secretary of State; senator; governor of Virginia; served on diplomatic missions resulting in the Louisiana Purchase and the acquisition of Florida; as fifth President of the United States he set forth the principles known as the *Monroe Doctrine*, ie., that the United States should not interfere in European affairs and that European countries should not interfere in the affairs of independent countries in the Western Hemisphere
1774   Francis Baily. Astronomer
1825   Herman Baumgarten. Writer
1835   Clarence Gordon. Author
1837   Fannie Hawkins. Pioneer
1852   Morgan O'Brien. Justice of New York Supreme Court

1866	Irma Grivot of Beaune (Sister Mary Ermine of Jesus) martyred in China in 1900
1869	Clement Young. Governor of California
1869	Bertram Goodhue. Architect
1871	Julius Finn. Champion chess player
1871	Louise Homer of the Met
1873	Harold Baver. Pianist
1878	Lionel Barrymore
1882	Henry Bellaman. Wrote *King's Row*
1889	Antonio de Oliveira Salazar. Statesman of Portugal
1898	Vincente Alexandre. Winner of Nobel Prize
1900	Jan Oort. Astronomer
1916	Furrucci Lamborghini. Industrialist; builder of farm machinery and motor cars
1917	Robert Anderson. Writer
1919	Fr. John Bernard McKenna. Famous G. I. of World War II; disciple of Padre Pio
1922	Alistair McLean. Novelist
1926	Harper Lee. Author

April 29

1544	Jean de la Bariero
1658	Giovanni Francesco Barbarigo. Scholar
1745	Oliver Ellsworth. Chief Justice of Supreme Court
1757	Reuben Tisdale of Col. Daggett's Regiment
1819	Joseph Larned. Inventor
1820	Henry Watkins Allen. Confederat officer; governor of Louisiana
1841	Edward Sill. Poet
1848	Thomas Campbell. Author
1854	Jules Henri Poincaré. Scientist
1860	Lorado Taft. Sculptor
1879	Sir Thomas Beecham. Conductor
1885	Robert Garland. Actor
1885	Adm. Frank Jack Fletcher. Task force commander at the Coral Sea, Midway and Guadalcanal
1893	Harold Urey. Novel Prize winner
1899	Duke Ellington
1909	Tom Ewell. Scientist
1926	Elmer Kelton. Writer

1931	Bill Lancaster. Legislator
1933	Keith Baxter. Actor
1936	Zubin Mehta
1936	Laura Marlene Thomas. Artist
1943	Duane Allen. Singer
1970	Uma Thurman. Actress
1970	Schuyler Grant of Broadway

April 30

1320	Casimir the Great. King of Poland
1504	Francesco Primaticcio. Painter
1651	St. John de la Salle
1654	Pierre Constant. Scholar
1740	Sarah Tisdale of Massachusetts Bay
1770	David Thompson. Explorer
1771	Hosea Ballou. Editor
1811	David Chase
1836	Robert Green
1850	Richard Zeckmer. Musician
1870	Franz Lebar. Composer
1872	Adm. Yates Stirling, Jr. President of Naval Examining Board; author
1878	Arthur Kennedy. Publisher
1885	Francis Bowes Sayre. Assistant Secretary of State
1891	James Edward Walsh. Missionary; first bishop of Chiangmen
1903	Fulton Lewis, Jr.
1910	Winfield Scott. Poet
1916	Claude Shannon. Mathematician
1920	Duman Hamilton. Race car driver
1933	Willie Nelson
1941	Elizabeth Ashley
1944	Jill Clayburgh
1945	Mimi Farina
1946	Don Schallender. Sprint swimmer

May 1

1218	Rudolph I. First Habsburg monarch
1285	Sir Edward Fitz-Alan

1672	Joseph Addison. Writer
1745	Nathaniel Emmons
1749	Ephraim Littlefield
1751	Joseph Whipple
1751	Ezra Ripley of the Colonies
1757	John Downing. Patriot
1757	James Walker of the Revolutionary War
1763	Joseph Coddington
1764	Benjamin Latrobe. Early American architect; engineer; designer of water works and canals
1769	Duke of Wellington. Defeated Napoleon at Waterloo
1816	Henry Rector. Governor of Arkansas
1825	James Carberry. Bishop of Hamilton
1825	Johann Balmer. Mathematician
1839	Hilaire Chardonnet. Discovered rayon
1841	Marcus Zucker. Author
1852	"Calamity Jane" Burk
1862	Ida Eisenhower
1863	Augustine Schinner. First bishop of Spokane
1875	Harriet Quimby. Aviator
1895	Leo Sowerly. Composer
1896	Gen. Mark Clark
1909	Kate Smith
1916	Glenn Ford
1925	Scott Carpenter. Astronaut
1929	Sonny James
1939	Judy Collins
1960	Steve Cauthen. Jockey

May 2

907	Boris I. Tsar of Bulgaria
1638	Edward Bernard
1660	Alessandro Scarlotti. Composer
1728	Hezekiah Todd of the Colonies
1778	Nathan Banks. President of Wesleyan University
1779	John Galt. Writer
1830	George Palmer. Settler
1844	Lucinda Baumgartner. Pioneer
1844	Minerva Hammer. Pioneer
1859	Walker Fanton Sherwood. Treasurer of Broome County

1862	Jesse Lazear. Pioneer researcher for cure for yellow fever
1865	Clyde Fitch. Author of *Beau Brummell*
1871	Francis Duffy. Chaplain; editor; author
1875	Owen Roberts. Justice of Supreme Court
1885	Hedda Hopper
1887	Michael Bohnen. Baritone
1887	Eddie Collins of Baseball Hall of Fame
1889	Rhea Halderman
1895	Lorenz Hart. Composer
1902	Werner Finck. Comedian
1904	Bing Crosby
1920	Marjorie Kriz. Writer
1941	Tony Adamowicz. Race car driver
1948	Larry Gatlin

May 3

1455	John II. King of Portugal
1469	Nicolo Machiavelli. Author of *The Prince*
1514	Ven. Bartholomew of Braga
1706	Thankful Alden. Descendant of Priscilla and John Alden
1740	Allerton Cushman of the Colonies
1741	Peter Cushing. Patriot
1753	William Clingan. Fought in the Revolution
1758	Bernard Hale
1794	James Osgood Andrew
1798	Thomas Arnold. Drummer boy of War of 1812
1802	Henry Wouters. Historian
1815	Barbary Gingery
1817	Orris Phelps
1823	Joseph Fordice
1840	Otto Mears. Railroader; legislator
1843	Edmond Prendergast
1844	Richard Carte. Singer
1853	Edgar Howe. Writer
1859	Andy Adams. Writer
1869	Julia Arthur. Actress
1883	Frances Alda. Soprano
1902	Walter Slezak. Actor
1904	Red Ruffing of baseball
1906	Mary Astor. Actress

1912	May Sarton. Poet
1915	Richard Lippold. Sculptor
1921	Sugar Ray Robinson
1928	Dave Dudley. Singer
1933	James Brown. Vocalist

May 4

1622	Jean Valdes. Painter
1654	K'ang-hsi. Emperor of China
1715	Richard Graves
1738	Elisha Loveland. Patriot
1770	François Gerard. Artist
1774	Gen. Samuel Willard Bridgham
1777	Louis Thenard. Discovered hydrogen peroxide
1778	John Shaw. Poet
1793	William Reeves. Senator
1796	Horace Mann of Congress
1801	George Washington Bonoparte Towns. Governor of Georgia
1812	John Stevenson. Governor of Kentucky
1815	Josiah Belden. California pioneer; town of Belden named for him
1820	Julia Tyler. First Lady (second wife of President Tyler)
1836	Congressman Tyre York
1840	George Gray. Senator
1847	Winfield Burbin. Governor of Indiana
1853	Philander Knox. Attorney General; senator; Secretary of State
1855	Sam Witherspoon of the House of Representatives
1866	Representative Clarence Van Duzen
1870	James English. Governor of Connecticut
1875	John Blaine. Governor of Wisconsin
1881	Norman Aimsley. Actor
1886	Cardinal Francis Spellman
1912	Adm. Bernard Clarey. Director of Navy Program Planning
1912	Chick Hiroshima. Auto racer
1922	Eugenie Clark. Author
1928	Wolfgang von Trips. Race car driver; winner of Berlin Grand Prix
1930	Roberta Peters. Soprano
1951	Pete Adams of Cleveland Browns
1959	Randy Travis

May 5

- 1210 Afonso III. King of Portugal
- 1504 Stanislaus Hosius
- 1666 Thomas Brooks
- 1747 Leopold II. Holy Roman Emperor
- 1761 Noah Croaker
- 1775 Margaret Penrose
- 1800 Louis Hachette. Publisher
- 1819 Stanislaus Moniuszko. Composer
- 1828 Jobez Cook Pierson
- 1830 John Batterson Stetson. Hatter
- 1832 Hubert Bancroft. Historian
- 1841 Gurdon Trumball. Artist
- 1842 Henry Northrop
- 1846 Henryk Sienkiewicz. Novelist
- 1852 Pietro Gasparre
- 1854 Arvilla Adalaide Brown
- 1867 Elizabeth Seaman. Journalist
- 1872 Bertrand Conway. Author
- 1874 Burton Cook Stewart
- 1876 John Garstang. Archeologist
- 1877 Bertha May Blue
- 1879 Joseph Tumulty
- 1890 Christopher Morley. Writer
- 1903 James Beard of cookbook fame
- 1904 Gordon Richards. Jockey
- 1915 Alice Faye. Actress
- 1939 Giorgio Di Sant Angelo. Designer
- 1942 Tammy Wynette

May 6

- 1493 Girolamo Seripando
- 1501 Pope Marcellus II (Marcello Cervini)
- 1574 Pope Innocent X (Giambattista Pamfili)
- 1615 Francesco Bressani. Missionary
- 1635 Johann Joachim Becker. Economist
- 1713 Charles Batteaux. Writer
- 1741 Archibald Coy
- 1743 Thankful Williams

1775	Mary Martha Sherwood. Author
1809	Commodore James Madison Frailey
1811	Smilax Cheney
1813	Platt Smith. Railroader
1837	John Henry Simonson. Inventor of double-treadle grindstone
1840	John Raines of Congress
1856	Adm. Robert Percy. Arctic explorer
1858	Edward Butler
1859	Luis María Drago
1860	Frank Dempster Sherman. Poet
1870	A.P. Giannini. Founder of Bank of America
1880	Ernst Kirchner. Painter
1881	Gregoria Martinez. Dramatist
1890	Martin Quigley. Publisher
1889	Wanda Neff
1890	Charles Lockwood. Inspector General, author of *Down to the Sea in Subs*
1895	Rudolph Valentino
1904	Harry Martinson. Poet
1908	Nancy Hale. Novelist
1913	Carmen Cavallaro. Bandleader
1913	Stewart Granger
1915	Orson Wells
1915	Theodore White
1923	Elizabeth Sellars. Actress
1931	Willie Mays of baseball
1939	Margaret Drabble. Actress

May 7

1610	Joseph Pancet
1661	Thomas Olney
1679	John Butterworth of New England
1704	Karl Graun. Composer
1713	Alexis Claude Clairaut. Scientist
1750	Amon Beebe of the Revolutionary War
1759	Solomon Abbott. Patriot
1774	William Bainbridge. Captain of the "Constitution"
1806	William Ullathorne
1812	Robert Browning. Poet
1822	André Garin. Missionary

1826	Varina Davis. First Lady of the Confederacy
1831	Richard Shaw. Architect
1833	Johannes Brahms. Composer
1837	Gen. Charles Byrne
1840	Pëtr Ilich Tchaikovsky. Composer
1844	Lucien Fenton
1857	Frederick Tuckerman. Anatomist
1888	Fenton Johnson. Poet
1901	Gary Cooper
1901	Commodore Roland Smoot
1923	Anne Baxter
1924	Gale Robbins. Actress
1933	Johnny Unitas. Quarterback

May 8

1492	Andrea Alciato
1521	St. Peter Canisius
1639	Giovanni Baciccio
1661	Tabitha Leffingwell. Settler
1668	Alain Lesage. Novelist
1740	Giovanni Paisiello. Composer
1756	Theophilus Lord. Patriot
1767	William Coombes. Missionary
1786	St. John Vianney, the Cure d'Ars
1821	William H. Vanderbilt. President of New York Central
1829	Louis Gottschalk. Composer
1846	Oscar Hammerstein. Playwright
1863	John Elmsley. Writer
1884	Harry S. Truman. President of the United States
1885	Thomas Costain. Author
1893	Eddie Roush of baseball
1895	Edmund Wilson. Poet
1895	Archbishop Fulton Sheen. Popular radio personality; author of *Life is Worth Living* Series
1910	Mary Lou Williams. Jazz pianist
1915	John Archer. Actor
1920	Sloan Wilson. Author
1926	Don Rickles
1930	Gary Snyder. Author
1940	Rick Nelson

1942 Angel Cordero. Jockey
1945 Keith Jarrett. Pianist
1964 Melissa Gilbert. Actress

May 9

909 St. Adalgar of Bremen
1740 Giovanni Paisiello. Painter
1744 Benjamin Bidwell. Patriot
1775 Jacob Brown. General of War of 1812
1783 Alexander Ross. Fur trader; author
1789 Sarah McCurdy
1798 Augustin Bonetty. Writer
1800 John Brown of Harper's Ferry
1814 John Brougham. Actor
1816 Jacob Blickenderfer. Railroad builder
1865 Annie Johnston. Author
1865 Elizabeth Jordan. Journalist
1870 Harry Vardon. Winner of British Open
1871 Mary Bell Shugart
1873 Howard Carter. Discovered tomb of King Tutankhamen
1873 Anton Cermak. Mayor of Chicago
1882 Henry Kaiser. Industrialist
1896 Austin Clarke. Poet
1904 David MacDonald. Director
1909 James Hagerty. Journalist
1912 Pedro Armendariz. Actor
1914 Hank Snow. Singer
1920 Richard Adams. Writer
1928 Pancho Gonzales of tennis
1931 Vance Brand. Astronaut
1941 Ritchie Valens. Singer
1949 Billy Joel
1957 Fred Markham. Cyclist

May 10

1272 Bl. Bernard Tolomei
1673 Joseph Aubery. Missionary
1697 Jean Marie Leclair. Composer

1730	George Ross. Signer of the Declaration of Independence
1731	Victor Louis. Architect
1739	John Thomas Troy
1755	Robert Gray. Navigator
1758	Joshua Tuggle
1760	Florimond Rouger. Composer
1789	Jared Sparks. Historian
1800	Sarah Peter. Founder of Philadelphia School of Design
1803	Luther Gates of Half Moon
1806	Margaret Graham
1813	Samuel Gerry. Artist
1821	Balthasar Boncompagni. Mathematician
1823	John Sherman. Secretary of Treasury
1858	Frederick Zeck. Pianist; composer
1883	Johnny Arthur. Actor
1899	Fred Astair
1902	David O. Selznick
1909	Maybelle Carter of the Grand Ole Opry
1916	Mellon Babbitt. Musician
1929	Antonine Moillet. Playwright
1936	Gary Owens

May 11

1499	Jean de Baudricourt. Governor of Burgundy
1738	Ferdinand Van Sickles
1751	Ralph Earl. Painter
1798	Cesar Despretz
1806	Joseph Barker
1811	Chang and Eng Bunker. The original Siamese twins
1811	Elijah Glover
1817	Fanny Cerrito. Ballet dancer
1823	Mary Fitzgibbon. Founder of New York Foundling Hospital
1824	Jean Gerome. Artist
1827	Septimus Winner. Songwriter
1838	Walter Goodman. Artist
1844	John Westervelt Bush. Charter member of the Buffalo Club
1845	Rev. Benjamin Hunter Du Puy who served at Gettysburg, Scottsylvania and Cold Harbor
1847	Godefroid Kurth. Historian
1852	Charles Warren Fairbanks. Vice President of the United States

1854	Ottmar Mergenthaler. Inventor of linotype
1871	Frank Schlesinger. Astronomer
1884	Alma Gluck. Soprano
1888	Adm. Willis Lee. Commander, Battleships Pacific Fleet
1888	Irving Berlin
1894	Martha Graham
1894	William Still. Musician
1896	Mari Sandoz. Writer
1912	Phil Silvers
1917	Gordon Auchincloss. Writer

May 12

1490	Gustav I. King of Sweden
1670	Augustus II. King of Poland
1683	Hannah Alden. Early New England settler
1700	Luigi Vanvitelli. Architect
1759	Ebenezer Kent. Patriot
1806	Gen. James Shields
1809	Giuseppi Giusti. Poet
1812	Edward Lear. Writer
1815	Hugh Gallagher. Missionary; editor
1817	Louis Couturier
1824	Edward Allis. Industrialist
1828	Dante Gabriel Rossetti. Painter
1837	William Gross
1842	Jules Massenet. Composer
1845	Gabriel Foure. Composer
1850	Henry Cabot Lodge. Senator; majority leader
1862	Louise Kellogg. Historian
1910	Gordon Jenkins
1914	Howard K. Smith. Commentator
1921	Farley Mowat. Author
1922	Ray Salvadori. Race car driver. Winner of Silverstone trophy
1925	Yogi Berra
1929	Burt Bacharach. Pianist
1930	Patricia McCormick. Diver

May 13

1652	Jonathan Emery. Served in King Philip's War
1655	Pope Innocent XIII (Michelangelo de Conti)

1717	Queen Maria Theresa of Austria and Hungary
1735	John Barrow. Missionary
1750	Hugh Coffey
1767	John VI. King of Portugal
1791	Anna Lytle
1792	Pope Pius IX (Giovanni Mastoi-Ferretti)
1799	John Van Buren. Congressman
1827	Mary Devereaux Clarke. Writer
1840	Alphonse Daudet. Novelist
1842	Sir Arthur Sullivan of Gilbert & Sullivan
1847	Linda Gilbert
1854	Wilhelm Diekamp. Historian
1857	Ronald Ross. Author
1867	Frank Brangwyne. Painter
1882	Georges Braque. Painter
1886	Jean Untermeig. Singer
1911	Maxine Sullivan. Singer
1912	Gil Evans
1914	Joe Louis
1924	Walter Wiggins. Publisher
1927	Clive Barnes
1927	Archie Scott-Brown. Race car driver
1940	Bruce Chatwin. Novelist
1950	Steve Wonder

May 14

1316	Charles IV of Luxemburg
1553	Margaret of Valois
1666	Victor Amadeus. Duke of Savoy
1737	Col. Samuel Holden Parsons. Fought at Ticonderoga
1742	Henry Walrath
1752	Timothy Dwight. Chaplain during Revolutionary War; president of Yale College
1754	John Leland. Writer; composer; legislator
1763	Amariah Chase
1765	James Furnish
1812	Alphons Czibulka. Pianist
1814	Charles William Russell. Historian
1817	Alexander Kaufmann. Poet
1862	John Joseph Cardinal Glennan

1867	William "Big Bill" Thompson. Mayor of Chicago
1881	Ed Walsh of Baseball Hall of Fame
1882	Matthew Walsh. Chaplain of World War I. President of Notre Dame
1888	Remi Ceillier. Scholar
1890	Carl Backman. Congressman
1897	Sidney Becket. Musician
1897	Virgil Geddes. Poet
1913	Walter Terry. Dance Critic
1937	Richard Howser. Baseball manager
1950	Mark Blu. Actor
1963	Carlos Lopez

May 15

1567	Claudio Monteverdi
1665	Ebenezer Weekes
1730	Col. Amos Wilcox who responded to the Lexington Alarm
1752	Moses Tullis. Patriot
1758	Increase Bliffin of the Colonies
1773	Prince Klemens Metternich. Chancellor of Austria
1805	Michael Balfe. Composer
1806	Gaetano Bedini. Inspired founding of North American College
1813	Liberty Stone of the battle of Franklin
1818	Richard Epperson
1820	Florence Nightingale
1827	Arthur Tremayne of the Light Dragoons
1845	Annie Strassburger
1858	Emily Folger
1856	Frank Baum. Writer
1887	Edwin Muir. Poet
1890	Katharine Porter. Novelist
1891	Florence Means. Author
1892	Adm. Charles Rosendahl. First Chief of Naval Airship Training
1905	Joseph Cotton
1909	James Mason
1918	Eddy Arnold
1918	Joseph Wiseman. Actor
1936	Anna Marie Alberghetti
1937	Trini Lopez

May 16

- 1578 Everard Digby
- 1611 Pope Innocent XI (Benedetto Odescalchi)
- 1718 Maria Agnesi. Mathematician
- 1751 James Madison. President of the United States
- 1763 Louis Vanquelin. Discovered chromium
- 1773 Adam Seybert. Congressman
- 1801 William Seward. Governor of New York
- 1824 Levi Parsons Morton. Vice President of the United States
- 1836 Kale Reignolds. Actress
- 1837 William B. Storg. President of the Atchison, Topeka & Santa Fe Railway
- 1842 Elizabeth Reed. Author
- 1844 Charles Reinhart. Artist
- 1881 Edward Archez. Scholar
- 1882 Anne Elizabeth McCormick. Journalist
- 1882 Carleton Hayes. Historian
- 1886 Douglas Freeman
- 1892 Richard Tauber. Tenor
- 1898 Kenji Mizogochi. Director
- 1905 Robert Bates
- 1905 John Conway. Author
- 1909 Luigi Villoresi. Race car driver; winner of Albi Grand Prix 1938
- 1912 Studs Terkel
- 1913 Woody Herman
- 1914 Randall Jarrell. Poet
- 1918 Barry Atwater. Actor
- 1928 Billy Martin of baseball
- 1940 Laine Kazan. Singer
- 1955 Debra Winger. Actress

May 17

- 1490 Albert. Duke of Prussia
- 1604 Vincent Baron. Writer
- 1728 Joseph Willard. Patriot
- 1741 John Penn. Signer of the Declaration of Independence
- 1749 Edward Jenner. Scientist
- 1760 Conrad Beibelheimer of the Colonies
- 1761 Nathan Lull

1768	Thomas Mann Randolph. Governor of Virginia
1791	Richard Miles. First bishop of Nashville
1794	Anna Jameson. Writer
1808	Charles Vogel. Composer
1817	Edgar Wadhams. First bishop of Ogdensburg
1833	Frederick Sears. Astronomer
1866	Erik Satie. Composer
1866	Franz Feldman. Scholar
1873	Henri Barbusse. Writer
1873	Dorothy Richardson. Novelist
1886	King Alphonse XIII of Spain
1901	Werner Egk. Conductor; composer
1905	John Patrick. Writer
1918	Birgit Nilsson. Singer
1923	Peter Mennin. Composer
1936	Thomas Mattingly. Astronaut
1949	Charles Mathes. Writer
1956	Sugar Ray Leonard

May 18

1515	St. Felix of Cantalice
1682	Mary Jefferson
1692	Joseph Butler
1711	Ruggiero Boscovich. Astronomer
1738	Jerremiah Lord
1795	Celestino Cavedoni. Archeologist
1824	Gen. Frederick Egloffstein of Civil War
1830	Karl Goldmark. Composer
1834	Sheldon Jackson. Editor
1838	Senator Watson Squire
1869	Rupert. Prince of Bavaria
1873	Dime Silverman. Founder of *Variety*
1885	Enrico Dutra. President of Brazil
1892	Ezio Pinza
1897	Frank Copra
1898	Meri La. Dancer
1902	Meredith Willson. Composer
1907	Irene Hunt. Writer
1909	Fred Perry of tennis
1912	Perry Como

1914	Pierre Balmoin. Designer
1918	Boris Christoff
1919	Margot Fonteyn
1920	Pope John Paul II (Karol Wojtyla)
1932	Florence Adams. Writer
1937	Brooks Robinson of Baseball Hall of Fame
1941	Diane McBaine. Actress

May 19

1718	Jonathan Reed
1721	Roger Sherman. Patriot
1757	Anthony Davenport
1759	David Coy of the Colonies
1795	Johns Hopkins
1800	Sarah Peale. Painter
1805	Francesco Vico. Astronomer
1815	Gen. John Gross Bernard. Chief Engineer of the Army of the Potomac
1816	Eli Compton
1820	Margaret Preston. Poet
1849	William Hardee
1859	Nellie Melva. Soprano
1868	Samuel Blythe. Newspaperman
1886	Bernadotte Schmitt. Historian
1892	Christian Werner. Race car driver; winner of silver medal
1914	Thomas Darreby. Missionary
1917	Robert Gleason. Author
1930	Lorraine Hansberry. Playwright
1931	Mark Boxer. Cartoonist
1934	Jim Lahrer. Newscaster
1935	David Hartman
1946	John Waihee. Governor of Hawaii
1947	David Helfgott. Pianist
1952	Grace Jones. Singer

May 20

1444	St. Bernardine of Siena
1470	Pietro Bembo. Author

1663	William Bradford
1742	Wessel Vernooy. Patriot
1764	Gottfried Shadow. Sculptor
1768	Dolley Madison. First Lady
1797	Nathaniel Greene. Editor
1799	Honore de Balzac. Novelist
1799	Sophie Gage
1818	William George Fargo. Co-founder of stagecoach
1821	James Pierce
1821	Christian Stadler. Famous wagonmaker
1824	George Otley
1825	John Ballentine. Congressman
1825	William Norris
1850	Joseph Acklen. Congressman
1851	Rose Hawthorne Lathrop. Founded St. Rose's Free Home for Incurable Cancer
1862	Breck Parkman Trowbridge. Architect
1890	Allan Nevins. Historian
1895	Adele Rogers St. Johns
1908	James Stewart
1917	Elizabeth Ogilvie. Author
1926	Bob Sweikert. Race car driver
1945	Wally Herger, Jr. Legislator; congressman
1949	Nick Raball of Congress

May 21

1352	Sir Philip Darcy
1454	Ermolao the Younger
1471	Albrecht Durer. Artist
1527	Phillip II. King of Spain
1664	Guilio Alberoni. Statesman
1688	Alexander Pope. Poet
1726	Isaac Van Vranken
1796	Reverdy Johnson. Senator; attorney general
1801	Robert Francis Withers Allston of West Point
1836	Stephen Reiter
1843	Thomas Lenihan. Bishop of Cheyenne
1855	Emile Verhaeren. Poet
1865	Philip Hale
1872	Richard Bennett. Actor

- 1893   Adm. Laurence Toombs. Commander Naval Forces Europe
- 1902   Marcel Brewer. Architect
- 1903   Alfred Roth. Architect
- 1904   Fats Waller
- 1915   Harold Robbins. Author
- 1917   Raymond Burr. Actor
- 1917   Dennis Day. Singer
- 1923   Edward Hulbert. Railroader
- 1925   Peggy Cass. Actress
- 1926   Robert Creeley. Poet
- 1985   Frustacci Septuplets

May 22

- 1675   Ives Andre. Mathematician
- 1703   Thomas Danforth. First American pewter
- 1715   François Bernis. Statesman
- 1733   Hubert Robert. Painter
- 1745   Levi Currier
- 1772   Levi Walker
- 1786   Arthur Tappan
- 1813   Richard Wagner. Composer
- 1815   Antonio Aparisi y Guijarro. Poet
- 1844   Mary Cassat. Painter
- 1848   Herman Shubert. Mathematician
- 1854   Jacob Schurman. Author
- 1859   Sir Arthur Conan Doyle
- 1897   Robert Neumann. Writer
- 1901   Maurice Tobin. Governor of Massachusetts
- 1902   Marie Fish. Oceanographer
- 1905   Paul Kristeller. Scholar
- 1907   Sir Laurence Olivier. Actor
- 1921   Marshall Teague. Auto racing great
- 1925   Jean Tinguely. Sculptor
- 1925   James King. Singer
- 1934   Bobby Johns. Race car driver
- 1938   Frank Converse. Actor
- 1946   George Best of soccer

May 23

- 1734   Franz Mesmer. Originator of hypnosis
- 1749   Martin Wikoff. Patriot

1759	Silence Alger of the Colonies
1799	Thomas Hood. Poet
1810	Martin Spalding. Bishop of Baltimore
1810	Margaret Fuller. Writer
1820	James Buchanan Eads. Bridge builder
1822	Richard White. Shakespearean scholar
1824	Ambrose Burnside. Union general
1827	Milton Slocum Latham. Governor; senator
1830	Henry Moore Teller. Senator; Secretary of Interior
1841	Giovanni Turini. Sculptor
1847	Edward Washburn
1880	William Albers of National Food Advisory Council
1889	Mabel Willebrandt. Assistant Attorney General
1899	General Alfred Gruenther. Recipient of Distinguished Service Medal
1901	Edmund Rubbia. Composer
1903	George Stone. Actor
1910	Artie Shaw. Bandleader
1912	Jean Francaix. Composer
1925	Mac Wiseman. Singer
1933	Joan Collins. Actress
1940	Gerard Larrouse. Racing driver
1945	Misty Morgan. Violinist
1948	Buddy Alan. Singer
1954	Marvin Hagler. Boxer

May 24

1494	Jacopo Pontarmo. Painter
1670	Vitus Richler. Writer
1715	Robinson Williams
1750	Stephen Girard. Patriot
1751	Sarah Norton of the Colonies
1781	Louis François Dauprat. Musician
1793	Edward Hitchcock. President of Amherst College; professor of theology and geology; led the way for establishing the American Association for the Advancement of Science; writer; established that there was no conflict between geological findings and scriptural accounts of creation
1803	Charles Bonaparte
1816	Emanuel Leutze. Artist

1819　Queen Victoria
1829　Benjamin Ulmann. Painter
1839　Martha Mullen
1850　Henry Grady. Orator
1852　Maurice Egan. Author
1855　Arthur Pinero. Playwright
1863　George Bernard. Sculptor
1866　Jean Giraud. Historian
1878　Harry Emerson Fosdick
1878　Lillian Gilbrith. Consulting Engineer. Mother of twelve children, two of whom wrote *Cheaper by the Dozen*
1883　Elsa Maxwell
1887　Mary Madeleva. Poet
1899　Suzanne Lenglen. Tennis star
1902　Marie Rose Ferron
1914　Lilli Palmer. Actress
1934　Jane Byrne. Mayor of Chicago
1955　Rosanne Cash

May 25

1550　St. Camillus
1597　Veit Eberman
1616　Carlo Dolci. Painter
1630　Theodore Beschefer. Missionary
1751　Prince Fish
1759　Littleton Beauchamp
1792　Jeduthan Stone. Early settler
1803　Ralph Waldo Emerson
1844　Clarence Dutton. Geologist
1845　Rev. H.F. Fairbanks. Author
1852　William Muldoon. Wrestler
1858　Charles Nutting
1863　Camille Erlanger. Composer
1877　Michael Ahern
1886　Philip Murray. Labor leader
1887　Padre Pio
1898　Bennet Cerf
1898　Gene Tunney
1908　Theodore Roethke. Writer
1918　Claude Atkins. Actor

1919	Dorothy Sarnoff. Singer
1925	Jeanne Crain
1926	Miles Davis
1927	Robert Ludlum. Author
1929	Beverly Sills
1939	Dixie Carter
1940	Tony Southgate. Race car driver

May 26

1700	Nikolaus Zingendorf
1712	Mary Heacock
1723	Daniel Grant
1731	Bernard Franklin
1737	Sebastian Cabot
1738	Abigail Salisbury
1748	William Barton
1792	Anne Hall. Painter
1806	Henry Knox Thatcher. Port Admiral; Union officer
1817	Denis MacCarthy. Poet
1867	Queen Mary of England. Wife of King George V
1873	Anna Burr. Author
1877	Claude Tucker of Oak Park
1886	Al Jolson. Singer
1890	Adm. Aaron Merrill. Victor of the Battle of Empress Augusta Bay
1894	Francis Walter. Co-author of McCarran-Walter Act; chairman of House Committee of Unamerican Activities
1898	Ernest Bacon. Composer
1900	Paul Grosjean. Scholar
1907	John Wayne
1909	Helen Anderson. Ambassador
1911	Ben Alexander. Actor
1920	Peggy Lee. Singer
1949	Hank Williams, Jr.
1951	Sally Ride. Astronaut

May 27

1332	Ibn Khaldum. Historian
1601	Anthony Daniel. Missionary

1725	Samuel Wart. Founder of Rhode Island College (Brown University)
1738	Nathan Gorham. Statesman
1753	Joshua Chritchfield. Patriot
1756	Maximilian I. King of Bavaria
1794	Cornelius Vanderbilt
1799	Jacques Halévy. Composer
1823	Gen. John Gray Foster of Mexican and Civil War
1829	John Goode. Soliciter General
1836	Jay Gould. Railroad builder
1937	Wild Bill Hickok
1839	Alfred Horatio Belo. Co-founder of *Associated Press*
1853	Julian Ralph. Author
1871	Georges Rouault. Painter
1879	Lucille Watson. Actress
1884	Max Brod. Author
1894	Dashiell Hammett. Novelist
1907	Rachel Carson. Author
1911	Vincent Price. Actor
1911	Hubert Humphrey
1915	Herman Wouk. Writer
1924	Sarah Vaughn. Singer
1928	Theo Musgrove. Composer
1939	Lee Meriwether

May 28

1371	John the Fearless. Duke of Burgundy; led crusade
1423	Katherine Percy of England
1696	Giovanni Berli
1741	Jonathan Loder. Patriot
1759	William Pitt the Younger
1760	Alexandre Beauharnais
1776	Martha Starr Ballou. Pioneer
1779	Thomas More. Poet
1818	Gen. Pierre Beauregard of the Confederacy
1833	Thomas Galberry
1841	Giovanni Sgambati. Composer
1847	John Marshall Hamilton
1850	Frederic Maitland. Historian
1864	Thomas Daly. Journalist; author

1876	Katharine Blunt
1883	Riccardo Zandonai. Composer
1888	Jim Thorpe. All American
1908	Ian Fleming
1916	Walker Percy. Novelist; winner of National Book Award
1917	Ray Fox. Racing driver
1917	Marshall Reed. Actor
1919	May Swenson. Poet
1923	Gyorgy Ligeti. Composer
1927	Eddie Sachs. Winner of Indianapolis 100
1938	Jerry West of basketball
1944	Gladys Knight. Singer
1952	Pamela McCarthy. Editor

May 29

1538	Thomas Pounde
1630	Charles II. King of England
1657	Alexander Lovell
1678	Adolphus. Prince of Fulda
1736	Patrick Henry. Patriot
1746	Ebenezer Bradford
1769	Ven. Anna Maria Taiga
1825	Thomas Emmet. Writer
1832	Joel Benton. Poet
1845	Elizabeth Pringle. Author
1848	John Singenberger. Composer
1855	Sir David Bruce. Scientist
1860	Isaac Albeniz. Composer
1873	Elizabeth Morrow. Writer
1874	G.K. Chesterton. Author
1892	Frederick Faust. Poet
1894	Beatrice Lillie. Actress
1903	Bob Hope
1905	Sebastian Shaw. Actor
1912	Pamela Johnson. Novelist
1913	Iris Adrian. Actress
1917	John F. Kennedy. President of the United States
1922	Joe Weatherly. Motorycle champion; winner of the Rebel 300 and the National 500
1930	March Fong
1939	Al Unser. Race car driver

May 30

- 1220 Alexander Nevski. Grand Duke of Vladimis
- 1739 Isaac Lowell of the Colonies
- 1742 Henry Goodman. Patriot
- 1746 John Hen Livingston of the Revolution
- 1755 James Goyne. Patriot
- 1769 Ann Merry. Actress
- 1782 Jerreh Kendall
- 1794 Zilpah Polly Banister. Settler
- 1808 Felix Barbelin. Founded St. Josephs Hospital, Philadelphia, first president of St. Joseph's College
- 1817 Hermann Hagen. Zoologist
- 1835 Alfred Austin. Poet
- 1853 Andrew Zabriski. Author
- 1856 Michael Lavelle. Founder of Lavelle School for the Blind
- 1863 William Rouse. Writer
- 1875 Giovanni Gentile. Artist
- 1886 Frank Walker. Composer; legislator
- 1888 James Farley. Postmaster General
- 1890 Larry Barretto. Novelist
- 1891 Urban John Vehr. First archbishop of Denver
- 1896 Howard Hawks. Director
- 1900 Carlos Villanueva. Architect
- 1901 Cornelia Otis Skinner
- 1907 Clarence Issenman. Editor
- 1909 Benny Goodman. Bandleader
- 1915 Larry Kelley of football

May 31

- 1443 Lady Margaret Beaufort. Mother of King Henry VII
- 1469 Manuel I. King of Portugal
- 1684 Timothy Cutler
- 1699 Ichabod Pond. Early settler of Wrentham
- 1751 James McCubbin Lingan. Patriot
- 1810 Horatio Seymour. Governor of New York
- 1818 John Albion Andrew. A founder of the Free Soil Party
- 1819 Walt Whitman. Poet
- 1824 Jessie Fremont. Writer
- 1868 John Seely. Author

1868   Frances Isabel Ledyard of the Colony Club
1883   Edward Duff. Chief of Naval Chaplains
1883   Eugene O'Sullivan. Congressman
1889   Athene Seylor. Actress
1894   Fred Allen
1895   George Steward. Writer
1906   Albert Reeves. Congressman
1908   Don Ameche
1923   Ellsworth Kelly. Painter
1923   Prince Rainier III
1941   Johnny Paycheck
1943   Joe Namath
1961   Lea Thompson. Actress
1965   Brooke Shields

June 1

1300   Thomas. Earl of Norfolk
1575   Etienne Bauny. Scholar
1637   Jacques Marquette. Explorer of Great Lakes and the Mississippi River
1761   Alexander Chubb of the Colonies
1765   Johann Hug
1800   Caroline Gentz. Novelist
1801   Brigham Young
1804   Mikhail Glinka. Composer
1816   Philip Kearney. Union general of Civil War
1816   Col. John A. Turley. Mayor of Portsmouth
1825   John Hunt Morgan. Confederate scout
1833   John Harlan. Supreme Court Justice
1878   John Mosefield. Poet
1884   Arthur Pound. Historian
1898   Molly Picon. Actress
1901   John Van Deuton. Playwright
1908   Helen Mulford Thompson. Manager of New York Philharmonic
1915   Johnny Bond. Singer
1921   Nelson Riddle
1922   Joan Caulfield. Actress
1925   Andy Griffith
1930   Charles Rangel of Congress
1933   Alan Ameche of football

1934   Pat Boone
1939   Jerold Adams. Writer
1955   Lesley Abrams. Comedian

June 2

1435   Ambrogio Calepino. Lexicographer
1653   Bethia Marston of New England
1714   Camillo Almici. Scholar
1725   Nicholas Dorsey
1732   Martha Washington. First First Lady
1740   Marquis de Sade
1741   George Straw
1762   Daniel Lovejoy of the Colonies
1816   John Godfrey Sax. Poet
1826   Tim Page
1835   Pope St. Pius X (Giuseppe Sarto)
1839   J.W. Ramsey. Mayor of Crawfordsville
1840   Thomas Hardy. Novelist
1845   George Adelbert Perry of the battles of Gravelly Run, Five Forks and Sunderland
1854   Max Ruhner. Scientist
1857   Edward Elgar. Composer
1858   Harry Shelley. Composer
1861   Helen Taft. First Lady
1870   Lucius Petty
1888   Albert Bennett. Mathematician
1895   Edith Meyer. Author
1904   Johnny Weismuller
1940   Constantine II. King of Greece
1942   Gary Patrick Stone. Artist
1944   Marvin Hamlisch. Composer

June 3

1696   Joseph Feuchtmayer. Sculptor
1726   James Hutton. Scientist
1733   Demas Lindley of the Colonies
1751   Pierre Rousseau
1760   Capt. Peter Duponceau of Revolutionary army

1762	Reuben Gray. Patriot
1763	James Yarbraugh
1771	Sydney Smith. Writer
1803	Wilhelm Knabe. Piano builder
1808	Jefferson Davis. President of the Confederate States of America
1809	Johann Buerkle. First furrier and glovemaker of Cincinnati
1810	Effie Titsworth. Pioneer
1814	W.J. Krug. Sheriff of Montgomery county
1819	Thomas Ball. Sculptor
1841	James Meline. Assistant United States Treasurer
1844	Garrett Hobart. Vice President of the United States
1853	Irwin Russell. Poet
1865	King George V of England
1878	Barney Oldfield. Race car driver
1901	Maurice Evans. Actor
1906	Mildred Brady. Consumer advocate
1911	Paulette Goddard. Actress
1925	Mindru Katz. Pianist
1942	Curtis Mayfield
1944	Martha Clarke. Dancer

June 4

1430	Petrus Burrus
1648	Ann Sturtevant of Plymouth
1737	William Gunnison. Patriot
1738	King George III
1738	Benjamin Corning. American patriot
1744	Jeremy Belknap. Historian
1751	Anthony Vanderslice of the Revolution
1761	John David. Missionary
1777	Louis Thenard. Discoverer of hydrogen peroxide
1870	George Sanford of Hall of Fame
1874	John Francis O'Hern. Bishop of Rochester
1876	Martin Welsh. President of Acquinos College; author
1889	Beno Gutenberg. Seismologist
1903	Charles Monroe. Bandleader
1903	Joel Berglund. Singer
1908	Walworth Barbour. Ambassador
1912	Rosalind Russell
1917	Robert Merrill

1919   Robert Barry. Baritone
1920   Fedora Barhieri. Mezzo-soprano
1923   Jane Barker. Deputy mayor of San Mateo
1924   Dennis Weaver
1926   Colleen Dewhurst. Actress
1934   Betty Adkins. Legislator
1937   Freddy Fender

June 5

1265   Dante Alighieri. Author of *Divine Comedy*
1599   Diego Velozquez. Painter
1646   Elena Cornaro. Scholar
1656   Joseph Tournefort. Botanist
1723   Adam Smith. Economist
1741   John Langdon. Delegate to Constitutional Convention
1762   Joseph Alger
1762   Bushrod Washington. Supreme Court Justice; son of John Augustine Washington and Hannah Bushrod Washington; nephew of George Washington
1800   Homer Hopson
1804   Robert Schomburgh. Explorer
1806   William Tyler. First bishop of Hartford
1840   Louis Hasbrauck of Ogdensburg
1846   Samuel Gorman. Scientist
1853   Alfred True
1856   Maud Goodwin. Historian
1862   Ellsworth White
1872   Ladislas Lazaro of Congress
1887   Ruth Felton Benedict. Author
1897   Georges Riviere
1900   Dennis Gabor. Scientist
1902   Walter Plunkett. Costume designer for *Gone With the Wind*
1906   Leonard Palmer. Writer
1913   Conrad Marca-Relli. Artist
1919   Richard Scarry. Writer

June 6

1502   John III. King of Portugal
1553   Bernardino Baldi. Writer

1606 Pierre Corneille. Poet
1755 Nathan Hale. American patriot famous for his speech, "I only regret that I have but one life to give for my country"
1756 John Trumbull. Painter of pictures in the Capitol Rotunda; president of American Academy of Fine Arts
1758 Samuel Reidinger. Patriot
1804 Louis Godey. Publisher
1809 Timothy Shay Arthur. Writer
1817 Alexander Forbes of the Oxford Movement
1834 Baldwin Dahl. Composer
1840 Gen. William Francis Bartlett. Commander of the 52nd Regiment of the Civil War
1842 Steele MacKaye. Actor
1861 Joseph Merriwether Terrell. Legislator; governor of Georgia
1868 Robert Falcon Scott. Explorer
1875 John Biurker. Editor
1892 Parker Moon. Historian
1898 Ninette de Valois. Founder of Royal Ballet
1898 Walter Abel. Actor
1905 Lazzlo Halasz. Conductor
1907 Bill Dickey. Catcher
1915 Vincent Persichetti. Composer
1925 Maxine Kumin. Writer
1932 David Scott. Astronaut
1935 The Dalai Lama
1956 Bjorn Borg of tennis

June 7

1687 Armand de la Richardie. Missionery
1711 François Jacquier
1747 James Lune. Patriot
1752 Solomon Rice
1752 Rachel Battles
1752 Joseph Truesdale of the Revolution
1755 Butler Stonestreet. Patriot
1760 Amos Ingraham
1788 "Beau Brummel." English dandy and leader of fashion
1813 John Barnet
1823 John Meisenheimer
1828 Hanna Stacy

1839	Sarah Tenney. Author
1848	Paul Gauguin. Artist
1850	Joseph Sibbel
1874	Andrew O'Connor. Sculptor
1876	Paul Bellot. Architect
1879	Knud Rasmussen. Explorer
1888	Steve Philbin. All American
1889	Francis Duff. Founder of the Legion of Mary
1897	George Szell. Conductor of Cleveland Orchestra
1899	Elizabeth Bowen. Writer
1909	Jessica Tandy. Actress
1917	Gwendolyn Brooks. Poet
1934	Philippe Entremont. Pianist, conductor
1940	Tom Jones
1940	Susan Alcott. Writer; editor
1957	Lee Munson of baseball

June 8

1552	Gabriello Chiabrera. Poet
1625	Giovanni Cassini. Astronomer
1671	Tomaso Albinoni. Composer
1682	Mehitabel Levet
1724	Franz Maulpertsch. Painter
1747	Josin Rogers. Patriot
1749	Frances Plowden
1770	James Mercer Garnett. Writer
1780	William Crolly
1783	Thomas Sully. Painter
1810	Robert Schumann. Composer
1813	David Dixon Porter. Civil War hero
1814	Charles Reade. Writer
1821	Samuel Baker. Explorer
1829	John Everett Millais. Painter
1830	Henry Clay Trumbull. Writer
1869	Frank Lloyd Wright
1917	Byron White of Pittsburgh Steelers; Supreme Court Justice
1918	Robert Preston. Actor
1919	Frank Mundy. Race driver
1921	Sheila Ryan. Actor
1923	George Kirby. Comedian

1925	Barbara Bush. First Lady
1934	James McCurrach. Pro squash player
1936	James Darren. Actor

June 9

1589	John of St. Thomas
1640	Leopold I. Holy Roman Emperor
1669	Lt. Abraham Past of the Train Band
1672	Peter the Great
1755	Stephen Underdown
1791	John Payne. Writer
1794	George Barrett. Actor
1810	Otto Nicolai. Composer
1812	Johann Galle. Astronomer
1815	Johnson Hooper
1827	Francis Finch. Author of *The Blue and the Grey*
1829	William Joshua Allen. Congressman
1861	Bernard Henry Pennings
1863	Ricca Allen. Actress
1865	Carl Nielson. Composer
1891	Matthew Smith. Editor
1893	Cole Porter. Composer
1900	Fred Waring
1910	Bob Cummings
1912	Patricia Cone. Author
1914	David Dempsey. Author
1929	Robert Badham. Congressman
1930	Ben Abruzzo. Balloonist
1934	Jackee Wilson. Singer
1961	Michael Fox. Actor

June 10

1688	James Francis Edward. King James III
1721	Francis Ducrue. Missionary
1725	Thomas Covington
1736	Zibiak White. Mayflower descendant
1740	Burgess Longworth
1741	Joseph Warren. Revolutionary War hero

1755	Deacon Samuel Brooks of the Colonies
1755	Andrew Gregg. Senator
1802	James Ware Bradbury. Founded New England's first normal school
1811	Joseph Albree Gilmore. Governor of New Hampshire
1818	Clara Novello. Singer
1835	Rebecca Latimer Felton. Senator
1837	Cyprien Bautrais
1862	Caroline Curtes. Actress
1881	William Arnold. First Chief of Chaplains of the United States to hold the rank of major general
1887	Harry Byrd. Statesman
1894	Maurice Lavanaux. Editor
1901	Frederick Loewe. Co-author of *My Fair Lady*
1907	Fairfield Porter. Painter
1911	Terence Rattigan. Playwright
1913	Wilbur Cohen. First employee of Social Security Administration
1921	Prince Philip
1922	Judy Garland
1928	Maurice Sendak. Author
1929	James McDivitt. Astronaut
1941	Dave Walker. Race car driver

June 11

888	St. Rimbert
1564	Joseph Heinz
1624	Jean Duhamel
1644	Prudence Leaver
1654	Jean Gerbillon. Missionary
1738	Benjamin Blatchford
1741	Joseph Warren of the Battle of Bunker Hill
1760	David Waterman
1776	John Constable. Painter
1825	Bayard Taylor. Writer
1821	Alexander Latta. Developer of steam fire engine
1832	Lucy Pickens
1835	Hon. William P. Britton
1864	Richard Strauss
1876	Alfred Kroeber. Anthropologist
1879	Julia Classen. Mezzo-soprano

1880	Jeannette Rankin. Congresswoman
1883	Albrey Fitch. Victor at the Battle of the Coral Sea
1895	Jacques Brugnon of tennis
1903	Ernie Nevers. All American
1913	Vince Lombardi of football
1913	Rise Stevens. Singer
1922	Benjamin Capps. Writer
1924	William Styron. Writer
1929	George Garrett. Novelist
1936	Chad Everett. Actor
1939	Wilma Burgess. Singer
1939	Jackie Stewart. Race car driver
1956	Joe Montana of football

June 12

1668	Thomas Archer. Architect
1723	Nathan Alden. Pioneer
1747	Dr. Philip Thomas of Revolutionary War
1762	Isham Browder
1798	Samuel Cooper
1806	John Roebling. Builder of Niagara and East River Bridge
1810	William Ramsey. Congressman
1819	Charles Kingsley. Settler
1833	James Weaver of Congress
1835	Ethan Allen Hitchcock
1837	William Gross. Archbishop of Portland
1846	Louis Roty. Winner of Grand Prix
1852	Catharine Dorn
1859	Thomas James Walsh. Senator
1877	John Gannon. Bishop of Erie
1877	Adm. Thomas Hart. Asiatic Fleet Commander
1885	John Moses. Governor of Minnesota; senator
1891	Alfred Montgomery
1897	Anthony Eden
1909	Virginia McCall. Author
1924	George Bush. President of the United States
1928	Vic Damon. Singer
1930	Innes Ireland. Race car driver
1943	Marv Albert. Sportscaster

June 13

- 40 Gnaeus Julius Agricola. Roman general
- 823 Charles II. King of Franks
- 1585 Antonio Ruiz. Pioneer
- 1631 Sarah Chauncey of England
- 1743 Francis Dana of the Colonies
- 1752 Fanny Burney. Writer
- 1803 George Carrell. First bishop of Covington
- 1805 John Henni. First archbishop of Milwaukee
- 1842 Camilla Urso. Violinist
- 1847 William O'Brien Pardow
- 1854 Bradley Fiske. Inventor
- 1865 William Butler Yeats. Poet
- 1875 Miriam Amanda (Ma) Ferguson. Governor
- 1880 Joseph Stella. Painter
- 1881 Lois Weber. Singer
- 1884 Étienne Gilson. Scholar
- 1890 Henry Shannon. Actor
- 1892 Basil Rathbone
- 1893 Dorothy Sayers. Mystery writer
- 1896 John Lavinos. Publisher
- 1897 Paavo Nurmi of track
- 1902 John Mussio. First bishop of Steubenville
- 1903 Red Grange. All American
- 1911 Luis Alvarez
- 1915 Don Budge of tennis
- 1916 Shirl Conway. Actress

June 14

- 1303 St. Bridget of Sweden
- 1716 Peter Harrison. Architect
- 1723 Johann Mayer. Composer
- 1761 Jesse Williams of the Colonies
- 1799 Patience Wolfe
- 1805 Bl. Benilda
- 1812 Fernando Wood. Mayor of New York
- 1816 Priscilla Tyler. Actress
- 1832 Micah Hayward Stone
- 1832 Maggie Mitchell. Actress

1851	John Zahm. Author
1862	John Cardinal Glennon. Archbishop of St. Louis
1874	Edward Bowes. Entertainer
1883	Marie Bernadot. Publisher
1884	John McCormack. Singer
1889	Knut Lundmark. Astronomer
1894	Grand Duchess Marie Adelaide of Luxemburg
1904	Thomas Carey
1907	Nicholas Bentley. Actor
1922	Kevin Roche. Architect
1933	Jerzy Kosinski. Novelist
1941	Michael Rogers. Legislator
1958	Eric Heiden. Speed skater; winner of Gold Medal
1969	Steffi Graf of tennis

June 15

1330	Prince Edward. Father of Richard II
1359	Robert Poyntz. Lord of Iron Acton
1636	Charles de Lafosse. Painter
1749	George Vogler. Composer
1758	John Thayer. Missionary
1767	Rachel Jackson. First Lady
1775	Carlo Porta. Poet
1797	Egbert Reasoner
1805	William Ogden
1809	François Xavier Garneau. Author
1810	Moses Thompson. Pioneer
1812	William Harrison Randall of Congress
1825	Charles Henry "Bill Arp" Smith. Journalist
1835	Adah Menken. Actress
1836	George Shoup. First governor of Idaho
1861	Erestine Schumann. Opera singer
1864	William Carter. Legislator; congressman
1884	Harry Langdon. Actor
1892	Minerva Letton. Artist
1898	James Lindsay Almond. Congressman
1905	Marie Estelle Hogue. Confectionary specialist
1910	Anne Freemantle. Historian
1910	David Ross. Composer
1937	Waylon Jennings. Singer
1958	Wade Boggs of baseball

June 16

- 1624 Maj. William Bradford. Son of Governor William Bradford; Chief military officer of the Colony
- 1637 Giovanni Colonna. Composer
- 1645 Mary Keyes
- 1736 Ichabod Cross
- 1743 Benjamin Davenport
- 1758 Abel Tower. Patriot
- 1762 Issac Abbott of the Colonies
- 1778 Charles Mercer. Founder of Chesapeake & Ohio Canal Company
- 1801 Julius Plucker. Scientist
- 1850 John Weissenburger
- 1858 Gustav V. King of Sweden
- 1869 Frederick O'Brien. Author
- 1875 John O'Rourke. Scholar
- 1890 Stan Laurel. Actor
- 1896 Dolores Schorsch. Author
- 1899 Helen Francesca Traubel. Soprano
- 1903 Alec Ulman. Formulated original book of racing rules
- 1905 Charles Morrow Wilson. Writer
- 1916 Hank Luisetti of basketball
- 1920 John Griffin. Author
- 1924 Arlene Hale. Author
- 1938 Judy Cannon. Actress
- 1950 Pat Nolan
- 1951 Robert Duran. Boxer

June 17

- 1239 Edward I. King of England
- 1657 Louis Duoin
- 1742 William Hooper. Signer of the Declaration of Independence
- 1749 Cheney Look
- 1788 Johann Achterfeld
- 1807 Adolphe Dechamps
- 1813 Thomas Silver. Inventor of grain-dryer and gas-burner for marine engines
- 1818 Charles Gounod. Composer
- 1823 Adams Reder

1844	Capt. W.P. Herron
1846	Phebe T. Eliza Clause
1867	Henry Lawson. Writer
1880	Carl Van Vechten. Novelist
1882	Chrystal Herne. Actress
1882	Igor Stravinsky. Composer
1883	Elbert Duncan Thomas. Missionary
1884	Prince Wilhelm. Duke of Sodermanland
1904	Ralph Bellamy. Actor
1908	Benjamin Morse. Author
1910	Red Foley. Singer
1917	Dean Martin
1923	"Crazylegs" Hirsch of Rams
1924	Narcisse Chamberlain. Editor; writer
1939	Frederick Vine. Scientist
1945	Eddie Marckx. Cyclist

June 18

1511	Bartolomino Ammannati. Sculptor
1691	Jacob Cooke
1716	Joseph Vien. Painter
1760	Friedrich Kaesemann. Indentured servant who lived to be over 100 years old
1806	Abijah Gilbert. Senator
1809	Adm. Sylvanus Godon of the attack on Port Royal
1856	Edward Genicot
1857	Henry Clay Folger. Founder of Folger Shakespeare Library
1870	Charles Baskerville
1889	David Lawrence. Governor of Pennsylvania
1896	Philip Barry. Playwright
1898	Carleton Hobbs. Actor
1900	Laura Hobson. Writer
1906	Kay Kyser
1906	George Pflaum. Publisher
1907	Jeannette MacDonald. Actress
1913	Sylvia Porter
1918	Robert Preston
1920	Jan Carmichael. Actor
1924	George Mikan of basketball
1937	Gail Godwin. Writer

1939   Lou Brock of Baseball Hall of Fame
1944   Jack Clancy of Green Bay Packers
1952   Carol Kane. Actress
1952   Isabella Rossellini. Actress

June 19

1566   James VI of Scotland
1623   Blaise Pascal. Mathematician
1665   Bl. Antonio Baldinucci
1717   Johann Stamitz. Composer
1748   George Wilson of the Colonies
1764   John Barrow. Writer
1774   Leonard Woods
1815   Cornelius Krieghoff. Painter
1829   Emma Handsaker
1834   Charles Spurgeon
1849   Sino Winser
1854   Alfred Catalani. Composer
1856   Ira Wood. Congressman
1858   John Osborne. Governor of Wyoming
1861   José Rizal. Poet
1861   Walter Nettleton. Painter
1881   Jimmy Walker. Mayor of New York
1882   Benedict Elder. Editor
1886   Dan Garvey. Governor of Arizona
1888   Anne Campbell. Writer
1902   Guy Lombardo
1903   Lou Gehrig
1919   Pauline Koel. Film Critic
1931   Toni Lander. Ballerina
1932   Pier Angeli. Actress

June 20

1271   Sir John Ferrers of Cardiff
1755   Casper Camp
1760   Richard Wellesley. Statesman
1773   Peter Early. Governor of Georgia
1796   Karolina Gerhardinger

1805	Constantino Brumidi. Painter
1813	Joseph Autran
1819	Jacques Offenbach. Composer
1820	Augustus Cadwell Beldon of Syracuse
1823	Jesse Reno of the Northern Virginia campaign
1836	Robert Curtis Ogden
1837	David Brewer. U.S. Supreme Court Justice
1853	Erich Schmidt. Historian
1860	Henry Reitz. Road Commissioner
1882	Sybil Arundale. Actress
1884	Francis Warren. Governor of Wyoming
1907	Lillian Hellman. Writer
1909	Erroll Flynn
1913	Juan III of Spain
1917	Jimmy Driftwood. Singer; fiddler
1921	Francisco Segura of tennis
1924	Chet Atkins. Singer
1924	Audie Murphy
1945	Anne Murray. Singer

June 21

1002	Pope St. Leo IX (Bruno of Egisheim)
1601	Godfrey Henschen
1639	Increase Mather
1646	Gottfried von Leibnitz. Mathematician
1709	Joseph Penrose. Settler
1766	Abraham Van Sciver
1771	Rebecca Hopson
1774	Daniel Tompkins. Vice President of the United States
1804	Johann Seidle. Poet
1805	Charles Thomas Jackson. Geologist
1834	Heinrich Weber. Historian
1836	W.J. Ott
1836	Sanford Bennet. Writer
1856	Hendrik Berlage. Artist
1857	William Byars. Writer
1859	Henry Tanner. Painter
1882	Rockwell Kent. Painter
1887	Norman Bowen. Scientist
1893	Alois Haba. Composer

1902	Adm. Harry Felt. Commander-in-chief United States Forces in Pacific and Far East
1913	William Masconi. Pocket billiard expert
1925	Maureen Stapleton. Actress
1959	Tom Chambers
1982	Prince William

June 22

1566	Sigismund III. King of Poland
1738	Jacques Delille
1757	George Vancouver. Explorer
1760	Joseph Colmar of the Colonies
1763	Etienne Mekus. Composer
1792	Bl. Dominic Barberi
1805	Ida Hahn. Author
1842	Lucetta Seaton
1846	Julian Hawthorne. Author
1856	Henry Haggard. Author of *King Solomon's Mines*
1856	Henry Tupper. Author
1879	John Dempsey. Secretary of Interior; governor of New Mexico
1883	John Bracken. Premier of Manitoba
1898	Erich Remarque. Race car driver
1900	Jennie Taurel. Opera singer
1903	Carl Hubbell of baseball
1906	Anne Morrow Lindbergh
1910	Katherine Dunham. Dancer
1915	Cornelius Wormerdam
1921	Barbara Vucanovich. Congresswoman
1921	Gower Champion
1943	Thomas Payne. Mayor of Bakersfield
1944	Klaus Brandauer. Winner of Golden Globe Award
1947	Pete Moravich of Basketball Hall of Fame

June 23

1666	Abiel Shurtleff. Settler
1683	Samuel Carpenter. First merchant of Philadelphia
1756	Gideon Castle
1750	Dieudonne Dalomieu. Mineralogist

1777	Frederick Bates. Governor of Missouri
1786	Nathaniel William Taylor
1807	James Harvey Pierson
1817	Louis Brisson
1810	Fanny Elssler. Ballerina
1819	Henry Gray. Artist
1827	Gen. Benjamin F. Coates. Senator; Deputy Director of Internal Revenue
1841	Wilhelm Waagen. Geologist
1859	William Bonney "Billy the Kid"
1875	Carl Miller. Sculptor
1876	Irvin Cobb
1881	Jesse Laney Boyd. Historian
1888	Francis Duffy. Judge; Senator; Vice-Commander, American Legion
1894	Wilber Brucker. Governor of Michigan; Secretary of the Army
1894	The Duke of Windsor
1912	Alan Turing. Inventor of Turing machine
1920	Donn Eisele. Astronaut
1929	June Carter. Singer
1940	Wilma Rudolph. Olympic track and field star
1943	James Levine. Pianist; conductor

June 24

1	St. John the Baptist. Last prophet
1198	Ferdinand III. King of Castile
1255	Roger de Somery
1311	Philippa of Hainaut
1475	Sir Thomas Stukeley of Kent
1499	Johann Brenz. Supporter of Augsberg Confession
1542	St. John of the Cross. Poet; author of the Spiritual Canticle, *The Ascent of Mt. Carmel, Dark Night of the Soul*
1770	Eli Ashmun. Senator
1777	Sir John Ross. Arctic explorer
1797	John Joseph Hughes. Archbishop of New York
1804	Stephan Endlicher. Botanist
1813	Henry Ward Beecher
1814	William Radford of Congress
1829	John Hogan
1830	Theodore Frelinghuysen Randolph. Governor of New Jersey
1831	Rebecca Davis. Novelist

1844	Clementine Deymann. Writer
1848	Brooks Adams. Historian
1854	Henry Dixon Allen. Congressman
1865	Robert Henri. Painter
1874	Selden Delaney. Author
1895	Jack Dempsey. Heavyweight champion
1895	Reid Railton. Auto racer
1911	Juan Fangi. World driving champion
1915	Fred Hale. Astronomer
1923	Jack Carter. Actor
1923	Yves Bonnefay. Poet
1931	Billy Casper. Winner of U.S. Open
1942	Michelle Lee. Actress
1946	Ellison Onizuka. Astronaut
1947	Mick Fleetwood of Fleetwood Mac

June 25

1625	Christian Herdtrick. Missionery
1675	Caleb Norton. Settler
1676	Moritz Hohenbaum
1730	Jonathan Truman of the Colonies
1734	Akinari Ueda. Writer
1737	Mary McFarland. Patriot
1759	William Plumer. Governor of New Hampshire
1761	Mathias Dague. Patriot
1786	Lucius Sargent. Poet
1801	Nancy Wilhoit
1806	Marcel Bouin. Author
1814	Gabriel Daubree. Geologist
1841	Thomas Jefferson Majors. Congressman
1825	William Crisman
1852	Antonio Gaudi. Architect
1860	Gustave Charpentier. Composer
1860	Valentine Rauh
1884	John Becker
1886	James Francis Cardinal McIntyre. Archbishop of Los Angeles
1887	George Abbott. Playwright
1900	Moses Hadas. Scholar
1903	George Orwell of *1984*
1907	Anne Revere. Actress

1924 Sidney Lunnet. Director
1925 Robert Venturi. Architect
1955 Lauie Ann Abramson. Choreographer

June 26

1462 Louis XII. King France
1716 Hendrick Lott
1725 Iqnaz Gunther. Sculptor
1730 Charles Messier. Astronomer
1741 John Langdon. Governor of New Hampshire
1742 Arthur Middleton. Signer of the Declaration of Independence
1791 James Fitz Randolph. Congressman
1828 Rebecca Isaacs. Actress
1834 John Alexander Anderson. Congressman
1846 William Bynum. Congressman
1869 Martin Nexo. Novelist
1875 Ricardo Stracciori. Baritone
1890 Adm. Oscar Badger. NATO Commander
1890 Jeane Eagels. Actress
1891 Sidney Howard. Playwright
1892 Pearl Buck
1898 Willy Messerschmitt. Designer of fighter bomber
1902 Antonio Brico. Pianist
1903 Bob Herman of baseball
1907 Robert McLaskey of Congress
1911 Bob Didrikson. Gold Medal Winner
1915 Charlotte Zolotow. Author
1928 William Jennings Sheffield. Governor of Alaska
1933 Claudio Abbado. Music Director of Vienna Philharmonic
1938 Neil Abercrombie. Congressman
1961 Greg LeMond. Cyclist

June 27

1350 Manuel II Palaeologus. Byzantine emperor
1801 Samuel Eccleston. Archbishop of Baltimore
1811 Eleanor Page Wood Walker
1818 James Breck. Missionary
1824 Daniel Smith. Settler
1829 Susan Partridge. Pioneer

1830	Jane Clark Spanger. Pioneer
1839	George Searle. Astronomer
1841	Alexander Baumgartner. Poet
1849	Francis Lathrop. Painter
1861	Fanny Davis. Pianist
1862	May Irvin. Actress
1864	Winnie Davis. Novelist; daughter of Jefferson Davis, known as the "Daughter of the Confederacy"
1867	Francis Oloffson
1872	Paul Dunbar. Novelist
1874	Anne Flexner. Playwright
1888	Antoinette Perry. Actress
1900	Fr. James Keller. Founder of "The Christophers"
1926	Frank O'Hara. Poet
1928	Jerome Ambro. Congressman
1928	Rudy Perpich. Governor of Minnesota
1941	William Brewster Ahern. Biologist
1946	George Gokel. Chemist; author; educator
1955	Isabella Adjani. Actress

June 28

1557	Ven. Philip Howard
1577	Peter Rubens. Painter
1490	Albert of Brandenburg
1491	Henry VIII
1534	Mattia Bellintani. Writer
1577	Peter Richens. Flemish baroque painter
1742	James Robertson. Pioneer; founder of frontier settlements in Tennessee
1760	Francis Dover
1801	Edward Barron. Irish missionary to Africa and United States
1814	Frederick Faber. Writer
1815	Robert Franz. Composer
1831	Joseph Joachim. Violinist; composer
1836	Lyman Gage. Secretary of Treasury
1843	John Morehouse
1844	John O'Reilly. Editor; author
1857	Emerson Hough. Writer
1887	Floyd Dell. Novelist, editor
1890	William Blandy

1891	Esther Faber. Winner of O'Henry Award
1894	Adm. Arthur Struble. Commander at Inchon landings
1902	Richard Rodgers of Rodgers & Hammerstein
1905	Ashley Montague. Writer
1906	Dr. Maria Goeppert Mayer. Nobel Prize winner
1926	Mel Brooks
1927	Patricia Thompson. Editor; publisher
1936	Theodore Wade. Publisher, author

June 29

1397	John II of Aragon
1680	Mary Noble. Pioneer
1721	Johann Kalb. Served at Valley Forge; killed at Camden
1760	Asa Lord of the Colonies
1797	Frederic Boroga. Missionary; composed the "Grammar and Dictionary of the Otchepee." First bishop of Sault Ste. Marie
1798	Georg Haring. Writer
1801	Pietro Alfiero. Composer
1827	Thomas Jefferson Eppes
1841	Adm. Arthur Burtis
1845	George Atkinson of Congress
1847	Ross Turner. Artist
1847	Patrick Millany. Author
1847	Patrick Azarias. Essayist
1852	Marianna Origer
1862	Gen. James McAndrews of Sioux campaign
1866	John Norris. Founder of Little Rock College
1868	George Ellery Hale. Astronomer
1871	Luisa Tetrazzini. Soprano
1872	Joseph Grace. Established Pan-Am
1877	Jessie Arnold. Actor
1893	Kossuth Purdy Chinn. Railroad developer
1901	Nelson Eddie
1914	Rafael Kuhelik. Composer; conductor
1921	Harry Schell. Race car driver
1945	Allan Albert. Producer

June 30

1470	Charles VIII. King of France
1685	John Gay. Dramatist

- 1717   John Gladding
- 1722   Georg Benda. Composer; violinist
- 1739   William Tucker. Patriot
- 1744   John Wertz
- 1754   George Gloninger. Patriot
- 1754   Stephen Tibbetts of the Colonies
- 1768   Elizabeth Monroe. First Lady
- 1789   Horace Varnet. Painter
- 1802   Benjamin Fitzpatrick. Governor of Alabama
- 1806   Philip Clark Masher of the Wabash and Erie Canal
- 1806   William Bradford Reed. Historian
- 1819   Lucile Grahn. Ballerina
- 1819   William Wheeler. Vice President of the United States
- 1826   Ozra Hadley. Governor of Arkansas
- 1845   Charles Knapp. Congressman
- 1855   Adm. William Caperton. Pacific Fleet Commander
- 1883   Adm. Royal Ingersoll. Atlantic Fleet Commander
- 1891   Leland Stanford Sedberry
- 1903   Robert Hannegan. Postmaster General; joint owner of the St. Louis Cardinals
- 1917   Lena Horne
- 1917   Beatrice Allen. Pianist
- 1917   Susan Hayward
- 1931   Roy Wagner. Editor

July 1

- 1506   Lajas II. King of Hungary
- 1725   Gen. Jean Rochambeau of Revolutionary War
- 1748   Abraham Longenecker
- 1750   Isaac Royce
- 1757   Hillebrant Lozier. Patriot
- 1781   James Bradley Finley
- 1831   Mary Olivia Nutting. Writer
- 1882   Susan Glaspell. Novelist
- 1882   Gen. William Cleary. Army chaplain; organized Chaplain's School
- 1887   Adm. Morton Deyo. Commander of Boston Naval Base
- 1892   James Cain. Author
- 1892   Jean Luecot. Painter
- 1899   Charles Laughton. Actor
- 1902   José Sert. Architect

1903	Dr. Gladys Anderson Emerson. Scientist
1912	Madeleine Stern. Author
1915	Jean Stafford. Novelist
1916	Olivia de Havilland. Actress
1916	Laurence Halprin. Designer
1925	Farley Granger
1931	Leslie Caron. Actress
1942	Karen Black. Actress
1949	Margaret Aoki. Actress
1952	Dan Aykroyd. Actor
1960	Evelyn King. Singer
1961	Carl Lewis. Track and field star
1966	Mike Tyson

July 2

419	Valentinian III. Roman Emperor of the West
1644	Abraham of Sancta Clara. Author
1714	Christopher Gluck. Composer
1742	Nathaniel Buell of the Revolutionary War
1745	Jess Lockwood. Patriot
1747	Christopher Slagle of the Colonies
1751	Peter Finn of the Revolution
1757	Daniel Dampier. Patriot
1808	Thomas Simpson. Arctic explorer
1810	Robert Toombs. Congressman; senator
1828	Joseph Unger. Writer
1846	Sarah Theresa Dunn. Missionary
1851	Elizabeth Burkey. Pioneer
1855	Bayard Tuckerman. Author
1861	Albert Ulman. Author
1871	Norman Duncan. Writer
1872	Cardinal Mundelein of Chicago
1873	Jeremiah Ford. Writer
1885	Lotte Lehman. Soprano
1888	James Boyd. Writer
1903	King Olav I of Norway
1908	Thurgood Marshall. Supreme Court Justice
1920	Isaac Stern. Violinist
1929	Imelda Marcos
1937	Richard Petty. Stock car driver

1947 Luci Baines Johnson
1955 Barbara Albright. Editor

July 3

1423 Louis XI. King of France
1567 Samuel de Champlain. Explorer
1731 Samuel Huntington. President of Continental Congress
1738 John Copley. Painter
1741 John Craven. Patriot
1742 Enos Chandler of the Colonies
1747 Henry Grattan. Orator
1781 Domenico Bentivoglio
1796 Maria Martin. Artist
1817 Alex Boyd. Early settler of Bureau County
1825 Erskine Nicol. Painter
1840 Jerusha Whitmarsh
1844 Dankmar Adler. Architect
1878 George M. Cohan. Actor
1883 Frank Kafka. Writer
1912 Elizabeth Coles Taylor. Novelist
1913 Dorothy Kilgallen. Journalist
1918 Theodore Apstein. Playwright
1920 Louise Albritton. Actress
1923 John Hartman. Singer
1935 Harrison H. Schmitt. Astronaut
1940 Cesar Tovar of major leagues
1943 Dennis Orrock. Mayor of San Buena
1953 Alynne Amkraut. Actress
1956 Debra Lee Babcock. Actress
1958 Gregory Aplin. Producer

July 4

1694 Louis Daquin. Organist
1714 Capt. Joseph Kidder
1729 George Webb of the Colonies
1753 Jean Pierre Blanchard. Balloonist; inventor of parachute
1784 Ebenezer Bass
1804 Nathaniel Hawthorne

1822	Henry Clemens Overstolz. Mayor of St. Louis
1826	Stephen Foster. Songwriter
1837	America Seaton
1838	Thomas Jefferson Auld
1857	Joseph Pennell. Artist
1863	John Gardner Coolidge
1868	Henrietta Leavitt. Astronomer
1879	Philippe Gaudert. Flutist
1872	Calvin Coolidge. President of the United States
1883	Rube Goldberg
1888	Henry Armetta. Actor
1895	Irving Caesar. Author of *No, No Nanette, Tea for Two*
1898	Gertrude Laurence
1900	Louis "Satchmo" Armstrong
1902	George Murphy. Actor; senator
1905	Lionel Trilling
1906	Vincent Schefer. Meteorologist
1913	Frances Innis. Sculptor
1928	Stephen Boyd. Actor
1929	Charles Tanner of Milwaukee Braves, Chicago Cubs and Cleveland Indians
1931	Rich Casares of Washington Redskins
1932	Gina Lollobrigida
1937	Sonja of Norway
1940	John Victor Anderson. Poet
1943	Geraldo Rivera
1943	Charles Wheeler. Pianist

July 5

1674	David Caulkins
1699	Ichabod Bryant
1744	Benjamin Vose. Patriot
1755	Sarah Siddons. Actress
1763	Ebenezer Seaver of Congress
1772	Samuel Polk of the War of 1812
1801	Adm. David Farragut
1805	William Hanna
1810	P.T. Barnum of the circus
1830	Joseph Hazen
1831	Jemina Hopps

1831	Marquis Bass
1842	Charles Henry Morgan of the Union Army; presidential elector
1842	Charles Bascom
1843	Mary Jane Kasheer
1852	Emma Shirk
1877	Wanda Landowska. Pianist
1878	Joseph Holbrooke. Composer
1880	Jan Kubelik. Violinist
1886	Felix Timmermanns. Author
1891	Jean Cacteau. Novelist
1898	Richard Condi. Conductor
1899	Marcel Arland. Writer
1937	Shirley Knight. Actress

July 6

1747	John Paul Jones of Revolutionary War who said, "I have just begun to fight."
1748	George Claghorn. Constructor of Old Ironsides (the "U.S.S. Constitution")
1757	William McKendree of the Continental Army
1759	Joshua Barney. Patriot
1759	William Lovejoy of the Colonies
1781	Thomas Raffles
1785	William Hooker. Botanist
1824	John Beveridge
1825	Randolph Rogers. Sculptor
1832	Maximilian. Emperor of Mexico
1854	Amzi Dixon. Editor
1859	Adam Byrd. Congressman
1865	Emile Dalcroze. Composer
1888	Annette Kellerman. Actress
1889	Adm. George Murray. Commander, Air Force Pacific Fleet
1995	Eleanor Clark. Writer
1899	Mignon Eberhart. Detective story writer
1915	Laverne Andrews of the Andrews Sisters
1917	Dorothy Kirsten. Opera singer
1925	Merv Griffin. Entertainer
1927	Janet Leigh. Actress
1928	Jim Rathman. Racing champion
1928	Patricia Stephenson. Artist
1946	Sylvester Stallone

July 7

1540	John Sigismund Zapolya. King of Hungary
1586	Thomas Hooker
1690	Placidus Bocken. Settler
1758	Jared Lockwood. Patriot
1810	George Sharsewood. Justice of Pennsylvania
1828	Heinrich Ferstel. Architect; known as the creator of modern Vienna
1836	Desire Girourd. Author
1838	Gustave Bickell. Pioneer
1838	Capt. Henry Seely of the "St. Louis"
1860	Gustav Mahler. Composer
1869	Percy Stafford Allen
1873	Jeremiah Ford. Editor
1884	Lion Feuchtwanger. Novelist
1884	Victor Salvatore. Sculptor
1887	Marc Chagall. Artist
1896	Clarence Manion of the Manion Forum; author; lecturer
1906	Satchel Paige of Baseball Hall of Fame
1911	Gian Carlo Menatti. Composer
1915	David Diamond. Composer
1915	Margaret Walker. Poet
1917	Elton Britt. Singer; guitarist
1924	Daniel Patrick O'Connell. Writer
1925	Charles White. Inventor; writer
1940	Richard Armay. Congressman
1948	Timothy Adams. Commentator

July 8

1478	Giangiorgio Trissini. Poet
1730	Gen. James Wadsworth of Revolutionary War
1747	Zackariah Whiting
1755	John Condit. Senator
1790	Fitz-Greene Halleck. Author of *Croaker Papers*
1819	Luigi Gregori. Artist
1829	David Turpi. Senator
1838	Count Ferdinand von Zeppelin. Military observer in Civil War attached to the Army of the Potomac; took part in expedition exploring the sources of the Mississippi River; made his first aerial ascent in balloon at Ft. Snelling, St. Paul, Minnesota

- 1841   Capt. Bruce Carr
- 1858   Frederick True. Zoologist
- 1869   William Moody. Writer
- 1872   John Bankhead. Senator
- 1875   Bernard McKenna. Author
- 1892   Richard Addington. Poet
- 1895   James McGranery. U.S. Attorney General
- 1900   George Antheil. Composer
- 1902   Salvatore Argenzio. Recipient of Colorado Authors' League award
- 1908   Nelson Rockefeller. Vice President of the United States
- 1910   Herman Lehmann. Scientist
- 1917   Faye Emerson. Actress
- 1946   Cynthia Gregory. Ballerina
- 1948   Kim Darby. Actress
- 1952   Jack Lambert of football
- 1958   Kevin Bacon. Actor

July 9

- 1578   Emperor Ferdinand II
- 1591   Jean Bagot. Publisher
- 1683   Philip V. King of Spain
- 1752   John Warne. Patriot
- 1755   Maj. Philip Worndle
- 1764   Ann Radcliffe. Writer
- 1810   Calvin Turner Fillmore
- 1811   Sarah Willis. Author
- 1814   Samuel Colt. Inventor of Colt Revolver
- 1815   Oran Roberts. Governor of Kansas
- 1819   Elias Howe. Developer of sewing machine
- 1829   Robert Armfield of Congress
- 1838   Florence Marryot. Novelist
- 1939   Karl Barmann. Concert pianist
- 1856   Nikola Tesla. Inventor of alternating current motor
- 1875   Edward Keating. Congressman; editor
- 1878   H.V. Kaltenborn. Broadcaster
- 1882   Richard Hageman. Musician
- 1888   Lycurgus Marshall. Congressman
- 1901   Barbara Cartland
- 1904   Kathryn Peck. Author
- 1905   Clarence Campbell. President of National Hockey League

1906	Elizabeth Lutyens. Composer
1928	Vince Edwards
1942	Richard Roundtree
1956	Tom Hanks
1976	Fred Savage

July 10

1333	Roger de Clifford. Sheriff of Cumberland
1451	James III of Scotland
1509	John Calvin
1710	Isaac Kingsland of Battle of Ticonderoga
1723	William Blackstone
1757	Joseph Ludlow. Patriot
1770	Experience Allen. Settler
1783	Bennet Tyler
1792	George Mifflin Dallas. Vice President of the United States
1815	Maj. Silas Battey
1817	William Gove. Editor; legislator
1829	Peter S. Kennedy
1834	James Whistler. Painter
1839	Adolphus Busch. Brewer
1849	Clement Gran. Artist
1857	Melora Shirk
1861	Timothy Corbett. First bishop of Crookston, Minnesota
1867	Peter Dunne. Journalist; writer
1870	Maurice Lugeon. Geologist
1875	Edmund Bentley. Writer
1882	Jewell Arthur Sperling. First judge of New Orleans
1883	Harry Allen. Actor
1885	Mary O'Hara. Novelist
1888	Hazel Abel of the Senate
1895	Carl Orff. Composer
1901	Adm. Daniel Gallery. Author of *Clear the Decks, Twenty Million Tons Under the Sea*
1920	David Brinkley
1923	Jean Kerr. Author
1932	Nick Adams. Actor
1933	Jan De Gaetano. Singer
1936	Ilana Lowengrub. Sculptor

July 11

- 1274 Robert I. King of Scotland
- 1603 Sir Kenelm Digby. Naval Commander
- 1744 Giovanni Devati
- 1764 James Slater
- 1767 John Quincy Adams. President of the United States
- 1771 Capt. John Rodgers. Hero of War of 1812
- 1819 Elizabeth Wetherell
- 1828 Reuben Peasley Cheney. Publisher of the *Green Mountain Kicker*
- 1836 Carlos Gomes. Composer
- 1838 John Wanamaker
- 1842 Henry Abbey. Poet
- 1846 Leon Bloy. Novelist
- 1847 John Henry Barrows
- 1856 Georgiana Drew Barrymore
- 1861 William Turney of Brewster County
- 1864 Charles Barr. Winner of Lipton Cup
- 1883 Lenora Templelien
- 1884 Francis Steck. Historian
- 1886 Walta Janka
- 1904 Bishop Alden Bell of Los Angeles and Sacramento; Air Force chaplain
- 1907 Tom Lea. Writer
- 1920 Yul Brynner
- 1938 John Ashton. Musician
- 1943 Rolf Stommeler. Race car driver

July 12

- 1303 Hugh de Courtenay. Earl of Devan
- 1468 Juan del Encina. Composer
- 1730 Josiah Wedgwood. Potter
- 1743 Jeremiah Wadsworth of Continental Congress
- 1746 Caleb Brokaw
- 1751 Bl. Julie Billiart. Foundress of the Congregation of the Sisters of Notre Dame of Namur
- 1762 Dominicus Lord
- 1817 Henry David Thoreau
- 1854 George Eastman. Founder of Eastman-Kodak

1865	Lucy Fitch Perkins. Writer
1870	Louis II. Prince of Monaco
1885	George Butterworth. Composer
1895	Kirsten Flagstad. Soprano
1895	Oscar Hammerstein. Songwriter
1896	Henlen Worden. Writer
1902	Elizabeth Montgomery. Author
1903	Marilyn Robertson. Soprano
1908	Milton Berle
1913	Willis Lamb. Nobel Prize winner
1920	Keith Andes. Actor
1922	Michael Ventris. Architect
1934	Van Cliburn. Pianist
1937	Bill Cosby
1941	Benny Parsons. Race car driver
1943	Christine Perfect McVie of Fleetwood Mac
1943	Bonnie Wheeler. Writer
1951	Michael McDonough. Artist

July 13

100BC	Julius Caesar
1590	Pope Clement X (Emilio Altieri)
1607	Wenceslaus Hollar. Etcher
1767	Joseph Speckbacher
1787	Count Pelligrino Rossi. Statesman; writer
1793	John Clare. Poet
1814	Joseph Alemany. Missionary; first archbishop of San Francisco
1815	James Seddon. Confederate Secretary of War
1820	John Garret. President of the Baltimore & Ohio
1839	William Tucker. Author
1841	Otto Wagner. Architect
1884	Adm. William Calhoun. Inspector General
1886	Fr. Edward Joseph Flanagan. Founder of Boys Town
1889	Queen Louise of Sweden
1898	Alexander Brook. Painter
1901	Mickey Walker of Boxing Hall of Fame
1914	Sam Hanks. Race car driver
1921	Charles Scribner, Jr. Publisher
1924	Carlo Bergonzi. Tenor
1928	Bob Crane. Actor

- 1933   David Malcolm Storey. Writer
- 1942   Harrison Ford
- 1945   Daniel Abramowicz of New Orleans Saints and San Francisco 49ers
- 1951   David Anthony

July 14

- 1486   Andrea del Sarlo. Artist
- 1634   Pasquier Quesnel
- 1750   Matthew Lyon. Congressman; Indian Agent
- 1756   Ebenezer Ballentine
- 1760   Thomas Linnen
- 1784   Jesse Elliott. Arranged the duel between Stephen Decatur and James Barron
- 1796   Jeremiah Stillwell
- 1803   Amaziah Richer
- 1811   Clara Fisher. Author
- 1818   Gen. Nathaniel Lyon. Killed in battle at Wilson Creek
- 1841   Washington Bowie
- 1857   Frederick Maytag. Mayor of Newton; developer of washing machine
- 1860   Owen Wister. Writer
- 1862   Florence Bascom. Geologist
- 1863   William Gauge
- 1874   Jess Tannebill of Cincinnatti Reds, Pittsburgh Pirates, New York Yankees, Boston Red Sox and Washington Senators
- 1833   Arthur James. Governor of Pennsylvania
- 1903   Irving Stone. Writer
- 1910   William Hanna. Cartoonist
- 1914   Gerald Ford. President of the United States
- 1917   William Eastlake. Novelist
- 1918   Ingmar Bergman
- 1930   Polly Bergen
- 1933   Del Reeves. Singer
- 1941   Julie Christie

July 15

- 1573   Inigo Jones. Architect
- 1606   Rembrandt von Rijn. Artist

1704	Augustus Spangenburg. Founder of first Moravian settlement in North America
1745	Moses Cushing. Patriot
1754	Amos Abbott, Jr. Patriot
1754	Jacob French. Composer
1781	Lisha Winter. Congressman
1799	Reuben Chapman. Governor of Alabama
1808	Edward Henry Cardinal Manning Archbishop of Westminster; established Oblates of St. Charles
1812	James Hope-Scott
1813	George Healy
1823	Ezra Hall Gillett. Author
1825	Joseph Abbott. Senator
1839	Representative John Seashoal Witcher
1841	William Whitney. Winner of English Derby; Secretary of Navy
1869	Gustave Robisher Tuska. Lecturer
1889	Marjorie Rambeau. Actress
1911	Marjorie Clark. Author
1919	Iris Murdock. Novelist
1921	Jack Beeson. Composer
1925	Carl Sanders. Governor of Georgia
1927	Ann Jellicoe. Actress
1938	Barry Goldwater, Jr.
1946	Linda Ronstadt
1954	Mario Kemper. World Cup winner
1960	Willie Ames. Actor

July 16

1194	St. Clare of Assisi
1496	Anne Fettiplace. Descendant of Charlamagne
1661	Pierre Iberville. Founder of Louisiana
1704	John Kay. Inventor of the flying shuttle
1723	Joshua Reynolds. Painter
1746	Giuseppi Piazzi. Astronomer; discovered Ceres
1750	Nicholas Darrow. Patriot
1759	Jerusha Kimball. Pioneer
1821	Mary Baker Eddy
1823	Gerson Wolf. Historian
1841	David Phelan. Author; editor
1851	Mildred Rutherford. Honorary president of the United Daughters of the Confederacy

1853   Federico Grate
1859   Jerusha Kimball. Pioneer
1863   Fannie Zeisler. Pianist
1868   Frederick Dawson. Pianist
1872   Roald Amendson. Explorer
1876   Edward Dent. Musician
1880   Kathleen Norris. Novelist
1883   Charles Sheeler. Painter
1887   Floyd Gibbons. Journalist
1906   James Still. Poet
1907   Barbara Stanwyck. Actress
1911   Ginger Rogers. Actress
1924   Bess Myerson

July 17

1698   Pierre Maupertius. Mathematician
1716   William Errington. Founder of Sedley Park School
1736   John Renne
1744   Elbridge Gerry. Vice President of the United States
1745   John Wentworth. Signer of Articles of Confederation
1745   Timothy Pickering. Statesman
1760   Alexander Maidonell. First bishop of Kingston, Ontario
1760   Otto Lummer. Scientist
1763   John Jacob Astor. Established fur trading post of Astoria
1764   Nehemiah Sampson. Patriot
1768   Fr. Stephen Badin. Missionary to Kentucky
1790   William Angel. Congressman
1797   Hippolyte Delaroche
1804   Karl Becker. Organist
1829   Sir Frederick Abel. Inventor of cordite
1831   Henry Biddle
1839   William Seymour of Albany
1846   De Alva Stanwood
1883   Goldsmith Bailey of Congress
1889   Erle Stanley Gardner. Creator of "Perry Mason"
1912   Art Linkletter
1917   Phyllis Diller
1935   Diahann Carroll
1951   Luci Arnaz

July 18

1552	Rudolph II. Holy Roman Emperor
1635	Robert Hooke. Discoverer of hairspring which is necessary for wristwatches
1656	Joachim Bouvet. Mathematician
1670	Giovanni Battista Bononcini. Composer
1720	Gilbert White. Writer
1750	Joseph Trafford
1757	Royall Tyler. Writer
1760	John Scott. Patriot
1782	Aristaces Azaria. Abbott; writer
1789	Thomas Carlin
1809	Augustin Backer. Bibliographer
1811	William Thackeray. Writer
1817	Peter Rathermel. Painter
1821	Pauline Viardot. Soprano
1833	Mary Agnes Tincker. Novelist
1837	Isadore Robot. Missionary
1852	Paul Carus. Editor
1853	Hendrik Lorentz. Nobel Prize winner
1858	William Redfield. Secretary of Commerce
1865	Laurence Houseman. Writer
1880	Elizabeth of the Trinity
1902	Chill Wills. Actor
1906	Clifford Odets. Playwright
1909	John Joseph Cardinal Wright
1913	Red Skelton
1916	Kenneth Armitage. Sculptor
1921	Senator John Glenn
1926	Elizabeth Jennings. Poet
1926	Margaret Laurence. Writer
1926	Buck Rainey. Writer

July 19

1729	William Ringgold of the Revolutionary War
1732	Richard Lucas. Patriot
1756	Caleb Logee. Fought for the Colonies
1775	John Shulze. Governor of Pennsylvania
1789	John Martin. Painter

1785	Mordecai Noah. Playwright
1788	William Rawle. Author
1794	James Marsh
1806	Josiah Gregg. Author
1816	Nehemiah Matson. Publisher; writer; established public library
1819	Robert Coffin. Writer
1820	Catherine Upham Draper
1827	Orville Plattof
1842	Frederic Greenholge. Governor of Massachusetts
1846	Edward Pickering. Astronomer
1860	Lizzie Borden
1865	Roy Beardsley
1868	Theodore Tuthill. Justice of New York Supreme Court
1891	Paul McNutt. Governor of Indiana
1905	Edgar Snow. Journalist
1916	Phil Cavaretta. Outfielder
1921	Elizabeth Spencer. Writer
1921	Dr. Rosalyn S. Yalow. Nobel Prize recipient
1931	Carol Brasnan. Musician
1937	George Hamilton IV

July 20

1661	Pierre Lemoyne Iberville. Explorer
1743	Pierre Denaul. Patriot
1745	Henry Holland. Architect
1747	Andrew Bearsticker of the Revolutionary War
1758	William Ferehee. Fought for the Colonies
1819	Samuel Griggs. Publisher
1820	Laura Keene. Actress
1823	Barney Williams. Actor
1836	Sir Thomas Clifford. Inventor of modern thermometer
1838	Augustin Daly. Playwright
1847	Max Lieberman
1864	Francis Walsh
1869	Joseph Byrnes. Congressman
1872	William Kramer of horse racing
1885	Robert Hogard. Author
1890	Courtney Savage. Author
1890	Theda Bara. Actress
1893	Alexander. King of Greece

1919	Edmund Hillary. Explorer
1934	Uwe Johnson. Novelist
1934	Henry Dumas. Poet
1953	Robin Haynes of Broadway
1963	Frank Whaley. Actor

July 21

1414	Pope Sixtus IV (Francesco Della Rovere)
1515	St. Philip Neri
1620	Jean Picard. Astronomer
1664	Matthew Prior. Poet
1812	Thomas Ashe. Congressman
1816	Paul Reuter of Reuter News
1818	Charles Robinson. Governor of Kansas
1821	Jonathan Wainwright of the Battle of Galveston
1832	Joseph Rainey of Congress
1835	Samuel Everett Stone
1847	Blanche Howard. Author
1851	Sam Bass
1860	Chauncey Olcott. Songwriter
1864	Frances Cleveland. First Lady
1865	William Ambrose Jones. First American bishop of San Juan
1896	Bourke Hickenlooper. Senator; governor of Iowa
1899	Ernest Hemingway
1902	Walter Farrell. Author of *Companion to the Summa*
1908	William Ezra Jenner. Senator
1920	Isaac Stern. Violinist
1926	Paul Burke. Actor
1934	Jonathan Miller. Writer
1946	Richard Hoxie. Actor
1951	Robin Williams
1954	Tia Riebling. Actress

July 22

1519	Pope Innocent IX (Giovanni Faccinetti)
1647	St. Margaret Mary Alacoque
1656	Caspar Moosbrugger. Architect
1680	Pierre Tencin. Statesman

1730	Daniel Carroll. Delegate to Continental Congress
1761	Ashbell Tillotson
1818	John Gregory Smith. Railroader; governor of Vermont
1824	John Dawson Shea. Historian
1830	Julia Dean. Actress
1839	David Moffat, Jr. Railroader; industrial developer of Colorado
1849	Emma Lazarus. Poet
1857	Frank Hamilton Cushing. Archeologist
1860	Mother Marie Butler. Founder of Marymount schools
1862	Evelyn Briggs Baldwin. Explorer
1880	Preserved Smith. Historian
1882	Edward Hopper. Painter
1890	Rose Kennedy
1898	Stephen Vincent Benet. Poet
1902	Rosamind du Jardin. Writer
1913	Licia Albanese. Soprano; debuted in "Madame Butterfly"
1918	Elizabeth Myers. Author
1920	Milton Marks
1949	Lasse Viren. Long distance runner

July 23

1730	Oliver Chatfield
1747	Faustino Arevalo
1748	David Boland of the Revolutionary War
1801	Robert Walker. Senator; Secretary of the Treasury
1809	Adam Prutsman. Pioneer
1810	David Enoch
1816	Charlotte Cushman. Actress
1824	Thomas Preston
1834	James Cardinal Gibbons. Archbishop of Baltimore; second cardinal of North America; author of *The Faith of Our Fathers*
1838	Samuel Byers. Author
1848	John Ezra Richards. Governor of Montana
1857	Albert Shaw. Writer
1857	William Cobb. Governor of Maine
1863	Samuel Kress. Retailer; art collector
1865	Charles Hiram Randall. Congressman
1867	Alfred Goither Allen of Congress
1867	Simion Rennewill. Governor of Connecticut

1874	"Sunny Jim" Fitzsimmons. The Grand Old Man of Thoroughbred Racing
1888	Raymond Chandler. Author
1892	Haile Selassie
1912	Michael Wilding
1926	Robert McCormick Adams. Anthropologist
1936	Anthony Kennedy. U.S. Supreme Court Justice
1936	Don Drysdale

July 24

1409	Alice Seymour
1686	Benedetto Marcello. Composer
1738	Elizabeth Becker. Novelist
1745	Sarah Campbell. Early settler
1756	Ebenezer Seeley
1759	Victor Emmanuel I
1783	Gen. Simon Bolivar
1786	Ichabod Bartlett. Congressman
1802	Alexandre Dumas. Writer
1803	Adolphe Adam. Composer
1804	Ira Aldridge. Actor
1815	William Groesbeck of Congress
1817	Adolphe. Duke of Nassau. Sovereign of the Grand Duchy of Luxembourg
1852	Charles Coster. Railroader
1853	William Gillette. Actor
1876	Jean Webster. Writer
1898	Amelia Earhart. Aviator
1916	John McDonal. Mystery writer
1921	Guiseppi di Stefano. Tenor
1922	Charles Mathias. Senator
1927	Wilfred Josephs. Composer
1931	Tony Hegbourne. Motorcycle and auto racer
1934	Willie Davis. Defensive end
1935	Dale Doig. Mayor of Fresno
1963	Julie Krone. Jockey

July 25

717	Elipandus of Toledo
975	Dietmar. Bishop of Mersebur; historian

1401	Jacqueline Hainaut. Countess of Holland
1421	Sir Henry Percy of the battle of Tawton Field
1512	Diego Covarruvias
1575	Christopher Scheiner. Astronomer
1607	Thomas Cheney
1654	Mercy Cooke
1718	Christopher Guice. Early settler
1732	Maj. Timothy Walker of the Rhode Island campaign
1750	Gen. Henry Knox. Founder of U.S. Military Academy
1751	Darius Seeger
1764	Daniel Abbot
1775	Anna Harrison. First Lady
1797	Janet McPherson
1801	Nancy Wilhoit
1805	Butler Denhem. Settler
1830	John Bausch. Developer of optics
1840	Sarah Humphrey
1844	Thomas Eakins. Painter
1853	David Belasco. Playwright
1859	Adm. Albert Niblack. President of International Hydrographic Bureau; author of *Why Wars Come*
1870	Denis McCarthy. Poet
1876	Louis Thomas McFadden. Co-author McFadden Pepper Act
1902	Frank Waters. Writer
1935	Tony Lanfranchi. Racer
1954	Walter Payton. Running back
1966	Holly Kay. Sprinter; fitness expert

July 26

1030	St. Stanislaus of Cracow. Bishop and martyr
1528	Diego Androda
1727	Horatio Gales of the Continental Army
1739	George Clinton. Vice President of the United States
1748	Nathaniel Bradford
1781	Commodore John Sloat
1782	John Field. Composer
1784	Commodore Charles Morris. Navy hero
1796	George Catlin. Artist
1805	Constantino Brunidi. Painter
1819	Eleanor C. Selders Epperson

- 1822 Mary Elizabeth Latimor. Author
- 1834 William Robinson. Writer
- 1852 David Henry Davidson
- 1866 Francesco Cilea. Composer
- 1870 Ignacio Zuloaga. Painter
- 1871 Patrick Healy. Historian
- 1875 Antonio Machado. Poet
- 1890 Adm. Daniel Callaghan. Killed at Battle of Quadalcanal
- 1893 George Grasz. Painter
- 1895 Gracie Allen
- 1897 Paul Gallico. Writer
- 1906 Mary Cover. Artist
- 1922 Jason Robards. Actor
- 1923 Wilhelm Hoyt. Pitcher
- 1928 Stanley Kubrick. Director
- 1956 Dorothy Hamill

July 27

- 1577 Diego Basalenque. Linquist
- 1648 Joseph Anthelmi. Historian
- 1660 Hannah Pond. Early settler of Wrentham
- 1742 Ezekiel Loring. Patriot
- 1752 Hodges Cutter of the Revolutionary War
- 1760 Justin Ashley of the Colonies
- 1796 Henry Lemke. Missionary
- 1801 George Airy. Astronomer
- 1820 Clement Vallandigham. Leader of the Copperheads
- 1835 Giosuè Carducci. Recipient of Nobel Prize
- 1867 Enrique Granados. Composer
- 1870 Hilaire Belloc. Noted author
- 1880 Joe Tinker of Baseball Hall of Fame
- 1894 Adm. Gerald Bogan. Served on the Saratoga, Lexington and Yorktown
- 1899 George Walberg of New York Giants
- 1900 Hans Haug. Composer
- 1906 Leo Durocher of baseball
- 1907 Ross Alexander. Actor
- 1916 Elizabeth Hardwick. Writer
- 1916 Keenan Wynn. Actor
- 1927 Jim Lodwick. Mayor of Denison

1939 Vernon Miller. Senator
1944 Bobbie Gentry. Singer
1948 Peggy Fleming

July 28

1609 Judith Leyster. Painter
1654 William Barrell. Early settler
1740 Joseph Da Costa
1746 Thomas Heywood, Jr. Signer of the Declaration of Independence
1751 Joseph Habersham. Member of Continental Congress; Postmaster General
1752 Artemas Cushman
1778 Commodore Charles Stewart
1811 Guilia Grisi. Soprano
1824 Consul Butterfield. Author
1844 Gerard Hopkins. Poet
1845 Etienne Bautraux
1859 Mary Antoinette Anderson. Actress
1866 Beatrix Potter. Author
1874 Alice Miller. Author
1879 Carrie Hall
1881 John Gresham Machen
1887 Tetsu Katayama. Premier of Japan
1893 Rued Longgaard. Composer
1901 Rudy Vallee. Actor
1902 Kenneth Fearing. Novelist; editor
1907 Graham Clark. Archeologist
1909 Malcolm Loury. Novelist
1928 Jackie O
1941 Ricardo Muti. Conductor
1943 Bill Bradley of basketball

July 29

1742 Isabella Graham. Philanthropist
1795 Edwin Stevens. Founder of Stevens Institute
1805 Alexis de Tocqueville. Historian
1805 Hiram Powell. Sculptor
1820 Clement Vallandigham. Commander of the Sons of Liberty; congressman

1824	Eastman Johnson. Painter
1861	Alice Roosevelt
1869	Booth Tarkington. Writer
1887	Sigmund Romberg. Pianist; composer
1892	William Powell. Actor
1905	Clara Bow. Actress
1905	Vivienne Bennet. Actress
1905	Stanley J. Kunitz. Poet
1907	Melvin Belli
1907	Robert Braidwood. Historian
1918	Edwin O'Connor
1918	Mary Lee Settle. Winner of National Book Award
1923	Richard Egan. Actor
1924	Luigi Musso. Race car driver
1925	Mikis Theodorakis. Composer
1926	Don Carter. Member, Bowling Hall of Fame
1930	Paul Taylor
1938	Peter Jennings
1941	David Warner. Actor
1972	Wil Wheaton. Actor

July 30

1511	Giorgio Asari. Painter
1540	Richard Fetherston. Martyr
1641	Regnier de Graaf. Anatomist
1643	John Badger
1704	Freelove Watson of Rhode Island
1738	Jeanne Latouche Cuyler of Revolutionary War
1739	James Claghorn. Patriot
1750	Experience Kinney
1806	Daniel Barringer. Congressman
1810	Aeneas Dawson
1812	Harrison Ludington. Governor of Wisconsin
1818	Emily Bronte. Novelist
1837	Gideon Parsons Nicholas
1849	Elma Rollins
1854	John Sharpe Williams. Congressman
1863	Henry Ford
1864	Adelbert Harrington
1868	Theodore Bauer

1888	Werner Jaeger. Author
1889	Julius Haldeman. Writer
1898	Henry Moore. Sculptor
1899	Archie Binns. Novelist
1899	Gerald Moore. Pianist
1924	Hugh Gallen. Governor of New Hampshire
1928	Johanna Albrecht. Singer
1941	Paul Anka. Singer
1947	Alan McCartney
1951	Mitchell Allen. Dancer
1957	Bill Cartwright of basketball
1958	Dailey Thompson. Decathlon winner

July 31

1526	Augustus. Elector of Saxony
1527	Maximilian II. Holy Roman Emperor
1702	Jean Attirete. Painter
1718	John Canton. First to observe magnetic storms
1749	Abraham Buford. Patriot
1750	Henry Tisdale. Minute Man
1802	Philander Page
1803	Sarah Maria Hewlett. Pioneer
1816	Gen. George Henry Thomas. Participated in Union victory
1820	John Work Garrett. President of Baltimore & Ohio Railroad
1824	George Miles. Writer
1828	Oskar Begas. Artist
1831	Paul Du Chaillu. Explorer; author of travel books
1851	Asbury Latimor. Senator
1861	George Barse. Painter
1863	Sidney Cotts. Governor of Florida
1864	John Kissell of Congress
1875	Jacques Villon. Painter
1904	Arthur Daley. Sportswriter; Pulitzer Prize recipient
1912	Milton Friedman. Economist
1916	Louise Smith. Race car driver
1919	William Quinn. Governor of Hawaii
1921	Whitney Young
1938	George McCord. Inventor of Spectophotometer
1940	James Nielson. State senator
1944	Geraldine Chaplin
1940	Congressman Wendell Bailey

August 1

- 10BC Tiberius Claudius I. Roman Emperor
- 1520 Sigismund II. King of Poland
- 1642 Ahmed II. Sultan of Ottoman Empire
- 1645 Eusebio Kino. Missionary; explorer
- 1711 Gideon Norton. Pioneer
- 1744 Jean Lamartine. Botanist
- 1757 James Devane. Patriot
- 1768 Karl Haller. Writer
- 1770 William Clark of Lewis & Clark Explorers
- 1775 John Wallis
- 1779 Francis Scott Key. Wrote lyrics of "Star Spangled Banner"
- 1786 William Salter. Commissioner
- 1815 Richard Dana. Author of *Two Years Before the Mast*
- 1818 Maria Mitchell. Astronomer
- 1819 Herman Melville. Author of *Moby Dick*
- 1824 Caspar Borgias. Bishop of Detroit
- 1839 Middleton Pope Barrow. Congressman
- 1873 Francis Sadlier. Publisher
- 1879 Ralph Hankinson of auto polo
- 1883 Lon Chaney. Actor
- 1888 Aline Kilmer. Poet
- 1908 Robert Dwyer. Archbishop of Portland
- 1921 Jack Kramer of tennis
- 1930 Geoffrey Halder. Dancer
- 1934 Bobby Isaac. Race car driver
- 1936 Yves Saint Laurent

August 2

- 1511 Andrew Barton
- 1627 Samuel van Hoogsbraeten. Dutch painter
- 1754 Pierre L'Enfant. Drew up the plans for the City of Washington
- 1757 John Rudisill. Patriot
- 1765 Louise Marie Baudouin. Founder of the Ursulines of Jesus
- 1802 Cardinal Wiseman. First archbishop of Westminster
- 1808 Augustus French
- 1809 Alexander Buchanan. Pioneer
- 1819 John Brooks. President of Omaha & Southwestern Railroad
- 1820 John Tyndall. Writer

1835	Elisha Grey. Inventor
1854	Francis Crawford. Novelist
1862	Duncan Scott. Poet
1866	Adrien de Gerlach. Leader of expedition to Antactica
1867	Ernest Dawson. Poet
1868	Constance I. King of Greece
1879	Rawghlie Stanford. Governor of Arizona
1882	Helen Martha Wright. Author
1891	Sir Arthur Bliss. Composer
1892	Jack Warner. Producer
1894	Westbrook Pegler
1896	Mitchell Hepburn
1898	Bl. Karolina Kozka. Martyr
1900	Helen Morgan. Singer
1910	Louis Zara. Novelist
1932	Peter O'Toole. Actor
1937	Sydney Mayer. Historian
1942	Louis Falco. Dancer

August 3

1651	Balthasar Permoser. Sculptor
1729	Richard Caswell. First governor of North Carolina
1752	Nicholas Coppernall
1836	Augustus Hopkins Strong
1836	Greene Black. Developed silver amalgam for dentistry
1839	George Goodale. Botanist
1859	George Lilley. Governor of Connecticut
1870	Maria Montessori. Educator
1872	King Haakon VII of Norway
1878	William O'Brien. Editor
1887	Rupert Brooks. Poet
1890	Frank Cobb. Boy Scout pioneer
1900	Ernie Pyle
1901	Cardinal Stefan Wyszynski
1903	Bernard Cullen. Director of Marquette League for American Indian Missions
1903	Walter Clark. Writer
1905	John Daly. Journalist
1905	Frances Frost. Novelist
1909	Adrieene Ames. Actress

1909	Walter Van Tilburg Clark. Writer
1921	Richard Adler. Composer
1924	Leon Uris. Writer
1926	Tony Bennett. Actor
1937	Diane Wakoski. Poet

August 4

1222	Sir Richard de Clare. Earl of Gloucester
1470	Bernardo Dovizi. Writer
1521	Pope Urban VII (Giambattista Castagna)
1604	François Aubignac. Poet
1721	Anselm Eckart
1756	Ezekial Longley. Patriot
1787	Armistead Mason. Senator
1792	Percy Shelley. Poet
1816	Louis Goesbriand. First bishop of Burlington
1816	Russell Sage. Congressman; railroad pioneer
1818	Gen. Lovell Harrison Rousseau of the Battle of Shiloh
1837	Princess Amamada
1841	William Hudson. Writer
1843	Elizabeth Hecktner
1845	Peter Cassidy
1875	Italo Montemezzi. Composer
1893	Adm. Russell Berkey. Commander of New York Navy Yard
1899	Ezra Taft Benson. Secretary of Agriculture
1900	Queen Elizabeth, the Queen Mother
1904	Achille Varzi. Winner of Tunis Grand Prix, Targa Florio and Monza
1908	Helen Kane. Actress
1908	Big Chief Feathers of Chicago Bears, Brooklyn Dodgers and Green Bay Packers
1910	William Shuman. Composer
1921	Maurie Richard of National Hockey League
1958	Mary Decker. World track and field record holder

August 5

1301	Edmund of Woodstock. Earl of Kent
1461	Alexander. King of Poland

1623	Marc Antonio Cesti. Composer
1665	Henry Stephens
1684	Leonardo Leo. Composer
1731	Jeremiah Gould. Patriot
1751	Drury Robertson of the Revolutionary War
1774	John Deveroux. First mayor of Utica
1802	Niels Abel
1811	Ambroise Thomas. Composer
1812	John Allison of Congress
1815	Edward Eyre. Explorer
1819	Gen. John Bidwell. Discovered gold in the Feather River
1821	Commodore Foreball Parker. Superintendent of Naval Academy; Chief Signal Officer; writer
1821	James Bedall McKean. Congressman
1868	Louis Brehier. Scholar
1876	Mary Beard. Author
1886	Bruce Barton. Author
1889	Conrad Potter. Winner, National Book Award
1906	John Huston. Actor
1914	David Brian. Actor
1930	Neil Armstrong. Astronaut
1940	Roman Gabriel. All American
1943	Leo Kinminen. Auto racer

August 6

1629	Samuel Blanchard of Hampshire
1651	François Fenelon
1665	Ebenezer Crane. Served in Sir William Phipp's expedition to Quebec; early settler of Braintree
1674	Jonathan Low
1675	Hon. Ezra Bourne
1736	Ignatius Lightner. Patriot
1739	Nathan Cutter of the Revolutionary War
1751	Sam Canby. Fought for the Colonies
1786	Ruth Nayes. Settler
1792	Elizabeth Mickle of Newburgh
1811	Judah Benjamin. U.S. Senator; Attorney General and Secretary of State of the Confederacy
1819	Samuel Carter. Admiral and general of the United States
1826	Julius Benedict. Early settler

1835	Henry Brownson. Author; scholar; chaplain in Union Army
1846	De Forest Richards. Governor of Wyoming
1852	Adm. Charles Badger. Superintendent of Naval Academy
1861	Edith Kermit Roosevelt
1870	Charles Rufus Brice. Legislator; judge
1876	Mary Zimbalist. Founder of Curtis Institute of Music
1881	Luella Parsons
1881	Lillian Albertson. Actress
1892	Ruth Suckow. Author
1894	Isaac Farrar of the Sons of American Revolution
1907	Gerald Dalrymple of Football Foundation Hall of Fame
1909	Karl Schnabel. Pianist
1911	Lucille Ball
1917	Robert Mitchum. Actor
1920	John Graves. Governor of Connecticult

August 7

317	Flavius Julius Constinius. Roman Emperor
1106	Henry IV. Holy Roman Emperor
1282	Elizabeth Plantagenet
1533	Alonzo de Ercilla. Poet
1656	Henry Basnage de Beauval. Author
1726	James Bowdoin. Revolutionary War leader
1742	Nathaniel Green. General of Continental Army
1760	Frederick Quitman. Missionary
1761	David Stephenson. Patriot
1764	Elihu Palmer. Settler
1774	Isaac Story. Poet
1779	Carl Ritter. Geographer
1788	Francis Shunk. Governor of Pennsylvania
1790	Jean Claude Colin. Pioneer
1795	Joseph Drake. Poet
1826	Samuel McLean. Congressman
1833	Senator Powell Clayton. Governor of Arkansas
1835	Roswell Flower. Governor of New York
1838	B.F. Waite. Pioneer
1843	Charles Warren Stoddart. Author
1843	Edwin Keightly. Congressman
1845	Martin Servies
1846	Francis Rawle of the American Bar Association

1847	Charles McCormack Reed
1852	Charles Wixom
1858	De Witt Clinton Badger. Congressman
1859	Austin Reed
1884	Reuben Wood of Congress
1888	Monty Wooley. Actor
1902	Ann Harding. Actress
1904	Ralphe Bunche
1907	Representative Harold Youngblood
1907	Suzanne Black
1926	Amory Houghton of Congress

August 8

1605	Cecil Calvert. Second Lord Baltimore
1799	Nathaniel Palmer. Discoverer
1833	Karl Decken. First to attempt to scale Mt. Kilimanjaro
1846	Samuel Milton Jones. Labor leader
1863	Florence Bailey. Author
1869	Post Wheeler. Poet
1876	Sen. Pat McCarran of McCarran-Walter Act
1877	Walter Bauer. Lexicographer
1882	Olga Samaroff. Concert pianist
1884	Sara Teasdale. Poet
1889	John Middleton. Author
1896	Marjorie Rowlings. Author
1902	Paul Dirac. Nobel Prize winner
1905	André Jolivet. Composer
1907	Jesse Stuart. Novelist
1918	Robert Aldrich. Director
1919	Dino de Laurenlis. Producer
1920	Loris Justice
1922	Jean Ritchie. Singer
1924	Rory Calhoun. Actor
1926	Richard Anderson. Actor
1929	Johnny Temple of Cincinnatti Reds, Cleveland Indians, Baltimore Orioles and Houston Astros
1932	Mel Tillis. Singer
1933	Connie Stevens
1937	Dustin Hoffman
1950	Keith Carradine. Actor

1955 Richard Alday. Composer
1988 Princess Beatrice of York

August 9

1541 Guy Lefevre. Poet
1653 John Oldham. Poet
1711 Hyacinthe Beaujeu
1735 Capt. John Kimball of Norwich
1739 Joseph Cutler. Patriot
1757 Elizabeth Schuyler (Mrs. Alexander) Hamilton. President of New York Orphan Society
1758 David Looney of Revolutionary War
1789 Solomon Juneau. Pioneer settler of Wisconsin
1799 George James. Novelist
1809 William Travis. Led defense of the Alamo
1811 Henry Faley. Historian
1828 Conrad Balanden. Novelist
1864 John Delaney. Founder and editor of *Guidon*
1869 Randle Ayrton. Actor
1874 Reynaldo Hahn. Composer
1884 Kenneth Latourette. Writer
1884 John McCain. World War II Navy aviation leader
1886 Pietro Yon. Composer
1894 Letitia Osborne. Novelist
1905 Leo Genn. Actor
1906 P.J. Travers. Writer
1925 Len Sutton. Race car driver
1927 Robert Shaw. Actor
1928 Bob Cousy of basketball
1931 Hurricane Jackson. Heavyweight contender
1944 Sam Elliott. Actor
1945 Ken Norton. Heavyweight
1952 Mark Andrews. Sportscaster
1957 Melanie Griffith. Actress

August 10

1019 Bl. Gundecar
1232 Sir John de Muscegras of Charlton

1333	Sir John de Ferrers of Chartley
1360	Francesco Zabarella
1397	Albert II. King of Germany
1753	Edmund Randolph. Governor of Virginia
1760	George Carnugh. Patriot
1809	John Kirk Townsend. Author
1813	William Henry Fry. Composer
1814	William Yancey. Senator of the Confederacy
1820	Enoch Lowe. Governor of Maryland
1821	Jay Cooke. Railroader; financier
1830	Lorenz Weix. Early settler of LeRoy, Wisconsin; served at the Battle of Nashville
1841	Mary Lathburg. Author
1842	Harriett Bass
1843	Charles Clark of the Battle of Santigo Bay
1843	Joseph McKenna. Supreme Court Justice
1845	Aaron Skinner. Astronomer
1856	Edward Doheny
1865	James Morrice. Painter
1874	James Forman Sloan. Jockey
1874	Herbert Hoover. President of the United States
1876	John Fitzpatrick. Historian
1886	Robert Atkins. Actor
1889	Norman Scott. Victor at Cape Esperance; killed at Quadalcanal
1893	Douglas Moore. Composer
1913	Jane Wyatt
1923	Rhonda Fleming
1927	Eddie Fisher

August 11

1542	Catherine Hastings. Wife of the Second Earl of Lincoln
1734	Miriam Norton. Settler
1738	John Bayard. Patriot
1757	John Dashiell of Revolutionary War
1762	Samuel Waldron
1764	Benjamin Sanders
1793	George Elder. Editor
1799	Joachim Barrande
1819	Martin Heade. Painter
1826	Andrew Jackson Davis. Author

1828	Edward Salomon. Governor of Wisconsin
1836	Charles Clark. President of New York, New Haven & Hartford Railroad
1836	Sarah Platt. Poet
1841	Zachariah Williams. Fought at Vicksburg, Little Rock, Mobile, Memphis, Chattanooga and the Red River
1841	Henry Honeychurch Garringe. Commander of the "Portsmouth"
1847	Benjamin Tillman. Senator; governor of South Carolina
1861	Anton Arenski. Conductor
1862	Carrie Bond. Composer
1865	Gifford Pinchot. Governor of Pennsylvania
1873	John Rosamond Jackson. Composer
1873	May Preston. Illustrator
1880	Theodore Shear. Archeologist
1897	Louise Bogan. Poet
1902	Lloyd Nolan. Actor
1902	I.W. Abel. Union leader
1921	Alex Haley. Author of *Roots*
1928	Arlene Dahl
1941	Raymond Mastrogiovanni. Sky marshall

August 12

1385	Thomas Ebendorfer. Statesman
1527	Renaud de Beaune
1591	Louise Le Gras. Foundress of the Sisters of Charity
1644	Heinrich Biber. Composer
1739	Nicholas Ware
1762	George IV. King of England
1774	Robert Southey. Writer
1781	Robert Mills. Architect
1794	Andrea Weix. Baker; has over 1,000 descendants
1806	Elizabeth Smith. Poet
1812	Samuel Haldeman. Scientist
1815	William Cassidy. Journalist
1852	Fr. Michael McGivney. Founder of the Knights of Columbus
1856	Diamond Jim Brady
1859	Katharine Bates. Author of "America the Beautiful"
1876	Mary Robert Rinehart
1880	Christy Mathewson of baseball
1881	Cecil B. DeMille

1882	George Bellows. Painter
1884	Arthur Aylesworth. Writer
1885	Helen Hunt Jackson. Author
1902	Marya Zaturenska. Poet
1925	Flannery O'Connor. Novelist
1927	Porter Wagoner. Singer
1929	Buck Owens
1933	Parnelli Jones. Race car driver
1954	Pat Methany. Musician

August 13

582	St. Arnulf. Mayor of the Palace; bishop of Metz
1422	William Caxton. Publisher
1751	Robert Thompson
1799	James Hunt. Congressman
1805	Robert Rantoul. Statesman
1814	Anders Angstrom. A founder of Spectroscopy
1815	Elizabeth Phelps. Writer
1818	Lucy Stone. Pioneer
1827	Francisco Amarim. Poet
1839	Michael Corrigan. Archbishop of New York
1849	Dwight Tryon. Painter
1860	Annie Oakley
1865	Emma Eames. Soprano
1866	Giovanni Agnelli. Founder of "Fiat"
1867	George Luks. Painter
1879	John Ireland. Composer
1890	Pauline Lord. Actress
1899	Alfred Hitchcock
1906	Chuck Carroll. All American
1908	Gene Raumond. Actor
1909	John Beal. Actor
1912	Ben Hogan
1921	Neville Brand. Actor
1943	Gary Hart. California state senator
1948	Kathleen Battle. Singer
1949	Bobby Clark of ice hockey

August 14

1552	Paolo Sarpi. Historian
1625	François de Harley

1729	Ignatius Loring of the Colonies
1734	Thomas Sumpter. Senator; Ft. Sumpter named for him
1740	Pope Pius VII (Barnaba Chiaramonti)
1742	George Nash. Governor of Ohio
1750	Charles Butler. Patriot
1791	Nathaniel Pierce
1805	N.W. Green. Pioneer tanner
1810	Peter Hasslacher
1860	Ernest Seton. Writer
1861	William McGarvey. Author
1867	John Galworthy. Writer
1871	Gilbert Garraghan
1880	Paul Foik. Historian
1878	Ethel Trout. Writer
1890	Ida May Boynton Wimberly of the Missouri Valley
1899	Thomas O'Connor
1926	Alice Adams. Writer
1936	Henry de Lumley. Archeologist
1940	Larry Waskinton. Composer
1942	Jack Oliver. Race car driver
1946	Susan St. James. Actress
1947	John Galsworthy. Writer

August 15

1195	St. Anthony of Padua
1233	St. Philip Benizi
1370	Elizabeth FitzRogers of Glen Manga Manor
1432	Luigi Pulci
1601	Johan Companius
1702	Francesca Zuccarelli. Painter
1752	Freeborn Garrettson of the Revolutionary War
1761	Edward Preble. U.S. Navy Commander of the Mediterranean Squadron in the War against Tripoli
1766	William Irving. Poet
1771	Sir Walter Scott
1785	Joseph Slaph
1787	Eliza Fallen. Writer
1809	Eugene Bore. Writer
1815	Sister Anthony O'Connell. Nurse in Civil War; called "The Florence Nightingale of America"

1822	Henry Maine. Historian
1830	Sir Henry Isaacs. Lord Mayor of London
1858	Emma Calve
1860	Florence Harding. First Lady
1864	Peter Yorke. Journalist; editor
1879	Ethel Barrymore
1883	Ivan Metrovic. Sculptor
1885	Edna Ferber. Writer
1889	Robert Benchley
1890	Jacques Ibert. Composer
1893	Harlow Herbert Curtice of General Motors
1904	Vincent Waters. Bishop of Raleigh
1905	Manfred von Brauchitsch. Race car driver
1910	Senator Thomas Kuchel
1912	Julia Child of cooking fame
1925	Oscar Peterson. Jazz pianist
1925	Mike Connors
1933	Lori Nelson. Actress
1950	Princess Ann
1992	Anne Lorraine of Fallon-Forreston

August 16

1355	Philippa Plantagenet. Wife of King Edward III
1397	King Albert of Hungary and Germany
1650	Hester Cooke. Early settler
1748	David Ziegler. Served under Washington and Steuben; commandant at Fort Harmar
1756	Oliver Wentworth. Patriot
1798	Mirabeau Lamar. Second president of the Republic of Texas
1815	St. John Bosco. Founder of the Salesians
1818	Thomas Wenthworth Peirce. President of Galveston & Houston Junction Railroad
1835	James Wilson. Congressman; Secretary of Agriculture
1862	Amos Alonzo Stagg. Coach; lived to age 103
1865	Dennis Cardinal Dougherty. Archbishop of Philadelphia
1877	Francis Murphy. Governor of New Hampshire
1883	Hugh Gibson. Ambassador to Belgium; author
1886	Hugo Gernshack. Fiction writer; inventor
1892	Sophie Braslau. Contralto
1894	George Meany. Labor leader

1904 Mary Taylor Hunt. Artist
1906 Prince Franz Joseph II of Lichtenstein
1913 Osie Hawkins. Baritone
1921 Ella Raines
1928 Ann Blyth
1933 Stuart Roosa. Astronaut
1943 Carolyn Adams. Dancer
1945 Suzanne Farrell. Ballerina
1947 John Howard. Olympic cyclist
1960 Timothy Hutton

August 17

1601 Pierre de Fermat. Mathematician
1629 John Sabieski. King of Poland
1768 Stephen Badin. Missionary
1785 Joseph Drake. Poet
1786 Davy Crockett
1794 Alexander Hohenlohe
1819 Joseph Salzmann. Missionary
1834 Pierre Senoit. Composer
1844 Menelik II. Emperor of Ethiopia
1864 Adm. Edward Eberle. Chief of Naval operations; author
1864 Gemma Bellincioni. Soprano
1872 Bessie Vannah
1887 Cardinal Samuel Stritch
1887 Karl Franz Joseph von Habsburg. Last emperor of Austria
1892 Mae West
1903 Abram Chasins. Concert pianist
1909 Eugene Willging. Editor; writer
1921 Maureen O'Hara. Actress
1925 John Hawks. Novelist
1923 Larry Rivers. Jazz musician
1929 James Wakatsuki. Speaker of Hawaiian House of Representatives
1930 Gerald Warren. Editor
1932 Red Kerr of basketball
1938 Francis McNamara. Race car designer
1952 Guillermo Vilos. Winner of U.S. Open

August 18

1243	Thomas of Capua. Composer, poet; cardinal, leading figure of his day
1276	Pope Adrian V (Ottobono Fieschi)
1587	Virginia Dare. First child born to English settlers in America
1657	Ven. Antonio De Jesus Margel. Missionary
1685	Brook Taylor. Mathematician
1735	François Xavier Feller. Author
1763	Solomon Blanchard. Patriot
1775	Meriwether Lewis. Explorer
1778	Gen. Ransom Noble
1803	Nathan Clifford. Took part in treaty with Mexico which made California part of the United States
1807	Charles Francis Adams. Historian; diplomat; congressman
1809	Philo Lewis Basset
1810	Constant Trayon. Artist
1830	Franz Joseph. Emperor of the Habsburg Empire. Reigned sixty-eight years
1846	Adm. Robley Evans
1849	Benjamin Godard. Composer
1852	Morton Plant. Winner of yachting races
1879	Gus Edwards. Wrote "By the Light of the Silvery Moon"
1885	Karl Adler. Bishop of Toledo; archbishop of Cincinnatti
1885	Basil Cameron. Conductor
1889	Jeseph McKee. Legislator; judge
1893	Burleigh Grimes of baseball
1898	Clemente Biondetti. Motorcycle, auto racer
1907	Howard Swanson. Composer
1907	Marguerita Zapoleon. Author
1918	Elsa Moranti. Novelist
1922	Shelley Winters
1925	Brian Aldiss. Science fiction writer
1927	Rosalyn Carter
1930	Johnny Preston. Singer
1933	Roman Polanski
1934	Robert Clems of Pittsburgh Pirates
1935	Rafer Johnson
1937	Robert Redford
1939	Molly Bee
1940	Bryan Bowers. Singer
1952	Patrick Swayze

August 19

- 1736 Constant Bosworth. Patriot
- 1742 Jean Dauberval. Dancer
- 1745 Johan Gahn. Scientist
- 1755 Ebenezer Mattson. Congressman
- 1778 William Gaston. Statesman
- 1785 Seth Thomas. Clockmaker
- 1793 Samuel Griswald. Author
- 1798 Charles Beck. Scholar
- 1836 Lathar Meyer. Scientist
- 1843 Charles Daughty. Poet
- 1846 Plencina Munson
- 1848 Gustave Caillebatte. Painter
- 1866 Sven Gronberger. Writer
- 1871 Florine Stettheimer. Artist
- 1871 Orville Wright of the "Wright Brothers." Inventors of the airplane
- 1881 Georges Enesco. Composer
- 1886 Robert Heger. Conductor; composer
- 1891 Milton Humason. Astronomer
- 1892 Chase Baromeo. Musician
- 1899 Bradley Tomlin. Painter
- 1900 Colleen Moore
- 1902 Ogden Nash
- 1903 James Cozzenn. Novelist
- 1905 Alphonse Clemens. Writer
- 1913 Eleanor Hull. Author
- 1915 Ring Lardner, Jr. Author
- 1931 Willie Shoemaker. Jockey
- 1933 Debra Paget. Actress
- 1940 Jill St. John
- 1954 Blaine Wesley Grant

August 20

- 1517 Antoine Granville
- 1561 Jacobi Peri. Composer
- 1632 Louis Bourdaloue. Orator
- 1644 Manuel Bernardes. Writer
- 1695 Duchess Marie Louise d'Orleans
- 1740 James Dashiell. Patriot

1745	Frances Asbury of the Revolutionary War
1787	John Niles. Postmaster General
1788	Roland Moseley of the Colonies
1793	Venus Ricker. Pioneer
1795	Commodore Robert Field Stockton. Governor and military commander in California; senator from New Jersey
1831	Eduard Suess. Geologer
1833	Benjamin Harrison. President of the United States
1843	Christine Nilsson. Soprano
1843	Edward Greene. Botanist
1859	Giovanni Grosoli
1862	John McGroarty. Poet; author
1910	Eero Saarinen. Designer of the St. Louis Gateway Arch
1912	Bob Sanson. Race car driver
1915	Vincent Edward Smith. Author
1922	Jacqueline Susann. Writer
1936	Clifford Andchor. Broadcaster
1936	Carla Fracci
1946	Connie Chung
1956	Joan Allen. Actress

August 21

1131	Baldwin II. Count of Edessa
1165	Philip II. King of France; crusader
1567	St. Francis de Sales
1609	Jean Ratrou. Playwright
1643	Afonso VI. King of Portugal
1721	Loring Cushing. Patriot
1725	Jean Greuze. Painter
1754	William Murdock
1783	Thomas Garrett. Abolitionist
1796	Asher Brown Durand. Painter
1805	August Bournonville. Choreographer
1808	William Gallagher. Poet
1811	William Kelly. Inventor
1821	Gen. William Barksdale of the Confederacy
1821	Joseph Dupont
1848	George Grenfell. Explorer
1854	Frank Munsey. Novelist
1874	James McGraw

1874	Gen. John Ryan. Commander of U.S. Forces in France and Belgium
1885	Francis Le Buffe. Author
1893	Lili Boulanger. Composer
1904	Count Basie
1913	Fred Agabashian. Race car driver
1930	Princess Margaret Rose of England
1936	Wilt Chamberlain

August 22

1358	Isabella of France
1607	Bartholomew Gosnold. Navigator
1647	Denis Papin. Inventor
1741	Jean La Perouse. Navigator
1744	John Cushing. Patriot
1774	Bartholomew Herder. Publisher
1797	Augustin Blanchet. First bishop of Walla Walla
1809	Henry Juncker. First bishop of Alton, Illinois
1834	Samuel Pierpont Langley. Astronomer
1835	Jason Ayres. City clerk; treasurer
1836	Archibald Willard. Painter; created the "Spirit of '76"
1841	Thomas De Witt
1841	Willard Glazier. Author
1852	George Meiklijohn. Assistant Secretary of War
1857	Georgia Cayvan. Actress
1862	Claude Debussy. Composer
1867	Charles Jenkins. Inventor
1868	Maud Powell. Violinist
1871	Butler Ames. Statesman
1881	Edward Johnson. Tenor
1891	Jacques Lipchitz. Sculptor
1893	Dorothy Parker. Recipient of O'Henry Award
1900	Elisabeth Bergner. Actress
1903	Addie Mae Hedrick. Author
1920	Ray Bradbury. Writer
1932	Gerald Carr. Astronaut

August 23

1579	Thomas Dempster. Author
1632	Rev. Isaac Chauncey of Harvard

1702	Magnon Le Roy of the Colonies
1754	Louis XVI of France
1785	Oliver Hazard Perry. Navy Hero. Said "We have met the enemy and they are ours"; victor at the Battle of Lake Erie in the War of 1812
1786	John Leeper of the Colonies
1796	Elizabeth Dangerfield Magill. Pioneer
1806	Johann Weiler. Founder of Belgium, Wisconsin
1814	James Roosevelt Bayley. Archbishop of Baltimore; established Seton Hall
1830	David Swing. Pioneer
1842	Osborne Reynolds. Scientist
1854	Moritz Moozkowski. Pianist; composer
1869	Edgar Masters. Poet
1884	Ogden Mills. Secretary of the Treasury
1900	Ernst Krenek. Composer
1912	Gene Kelly. Dancer
1913	Bob Crosby
1917	Tex Williams
1927	Allan Kaprow. Painter
1931	Hamilton Smith
1934	Barbara Eden
1938	Ronny Cox. Actor
1942	Patricia McBride. Actress

August 24

1113	Geoffrey V. Plantagenet; Duke of Normandy
1198	Alexander II. King of Scotland
1237	Sir William de Huntingfield
1429	Sir John Boteler of Bewsey
1603	Abu Bahadur of Khiva
1727	Eliphalet Adams. Patriot
1729	George Hay. Writer
1742	Lovell Bullock. Patriot
1760	Baxter Ragsdale of the Colonies
1764	John Dubois. Bishop of New York
1772	William I. King of the Netherlands
1823	Gen. John Newton of the battles of Gettysburg and Fredericksburg
1848	Kate Claxton. Actress
1863	May Sinclair

1878	Walter Eaton. Writer
1888	Loui Kucera. Bishop of Lincoln
1895	Richard Cardinal Cushing
1899	Jorge Borges. Writer
1903	Graham Sutherland. Painter
1904	Adm. Morton Mumma. P.T. Boat Commander
1930	Roger McCluskey. Race car driver
1936	Lowell Swalve. President National Pork Producers; President Farm Bureau
1942	Egon Madson. Dancer
1944	Gregory Jarvis. Astronaut
1960	Calvin Ripkin, Jr. of Baltimore Orioles

August 25

1252	Sir Hugh Poyntz of Somerset
1530	Ivan IV. First Tsar of Russia
1651	François Baert. Writer
1724	George Stubbs. Painter
1735	Arent Vedder. Patriot
1783	Capt. Samuel Reid of War of 1812
1792	Joseph Chandler. Journalist
1819	Allan Pinkerton. Detective
1822	George Wolff. Editor
1836	Bret Harte. Novelist
1850	John McShane. Congressman
1851	George Lathrop. Poet
1858	Silas Holcomb. Governor of Nebraska
1873	Blanche Bates
1879	Robert Nisbet. Painter
1895	Robert Hurley. Governor of Connecticutt
1900	Chuck Carney of National Football Hall of Fame
1901	Kjeld Abell. Playwright
1905	Clara Bow
1909	Michael Rennie. Actor
1915	Laurance Falco. Bishop of Amarillo
1918	Leonard Bernstein. Conductor
1919	George Wallace. Governor of Alabama
1921	Brian Moore. Writer

August 26

- 1662 Isaac Wright. Pioneer
- 1671 Supply Weeks. Settler
- 1688 Egidio Forcellini. Lexicography
- 1693 August Allen. Descendant of Priscilla and John
- 1719 Carlo Sebastiano Bernard. Scholar
- 1735 Consider Crapo. Hero of Revolutionary War
- 1740 Joseph Montgolfier. Inventor of balloon
- 1743 Antoine Lavoisier. Father of modern chemistry
- 1811 Ginery Twichell. Congressman
- 1819 Prince Albert. Husband of Queen Victoria
- 1825 Henry Ream. Pioneer
- 1830 J.P. White. Thresher builder
- 1838 John Wilkes Booth
- 1839 Hernando DeSoto Money. Senator
- 1848 Charles Stanton Devas. Writer
- 1850 John Borradaile. Hotelier
- 1852 Mary Chambers. Pioneer
- 1873 May Becker. Author
- 1874 Zona Gale. Novelist
- 1874 Joseph Dunn. Scholar
- 1875 John Buchan. Writer
- 1886 Rudolph Belling. Sculptor
- 1899 Rufino Tamayo. Artist
- 1901 Gen. Maxwell Taylor
- 1906 Albert Sabin. Developer of polio vaccine
- 1946 Swede Savage. Race driver

August 27

- 1300 Ralph Basset of Great Weldon
- 1512 Friedrich Staphylus
- 1637 Charles Calvert. Governor of Maryland Colony
- 1758 Cornelius Longendyke. Patriot
- 1763 Francis Brooke of Smithfield
- 1765 José Viader. Missionary
- 1772 Gideon Blackburn
- 1796 Sophie Smith. Founder of Smith College
- 1800 Gen. William Harney of the Oregon Territory
- 1840 Cyrean Battey

1851   Levin Perdue. Settler
1865   Gen. Charles Dawes. Federal Comptroller of the currency
1866   Julia Marlowe. Actress
1867   Umberto Giordano. Composer
1874   Carl Basch. Nobel Prize winner
1882   Sam Goldwyn. Producer
1884   Rose McClendon. Actress
1885   Daniel O'Connell. Editor
1888   Gen. Arthur McDermott. Served in Mexican campaign and World War I
1894   George Schuster
1899   C.S. Forester. Biographer
1904   Norah Lofts. Author
1908   Frank Leahy. Coach; Knight of Malta
1908   Lyndon Johnson. President of the United States
1910   Mother Teresa
1916   Keenan Wynn
1937   Tammy Sands
1943   Tuesday Weld

August 28

1658   Honore Townely. Scholar
1735   Andreas Bernstorff of Revolutionary War
1749   Johann Wolfgang Goethe. Poet
1760   James Dey. Patriot
1764   John Hassler of the Colonies
1774   St. Elizabeth Ann Seton
1831   Lucy Hayes (Mrs. Rutherford B. Hayes). First Lady; first president of the Woman's Home Missionary Society
1844   Caroline Boas Phillips. Pioneer
1845   Adm. "Shaky" Davis. Writer of scientific works
1849   Rufus Zogbaum. Artist
1871   Nelle Eberhart. Wrote *From the Land of the Sky-Blue Water*
1872   Ada Harrington
1873   Henry Althoff. Bishop of Belleville
1874   Moulton Hills
1885   Grover Gehant. Justice of the Peace; notary public
1896   Liam O'Flaherty. Writer
1894   Karl Bohm. Conductor
1897   Morris Ankrum. "Man with the X-Ray Eyes"

1899	Charles Boyer. Actor
1910	Morris Graves. Artist
1913	Robertson Davis
1913	Robert Irving. Conductor
1914	Richard Tucker. Tenor
1924	Janet Frame. Writer
1924	Peggy Ryan. Actress
1956	Scott Hamilton

August 29

1737	Philippe du Molette. Scholar
1748	James Armstrong. Congressman
1756	Benjamin Conant. Patriot
1759	William Beckford
1763	Benjamin Brink of the Revolution
1769	Ven. Rose Philippine Duchesne
1780	Jean Ingres. Painter
1780	Richard Rush. Secretary of State
1786	Gen. Charles Gratiot Cabanne. Hero of the Delaware River
1792	Charles Finney. Scholar; president of Oberlin College
1809	Oliver Wendell Holmes, Sr. Poet
1813	Thomas Allen. Congressman
1824	Eliza Starr. Author
1889	Patrick Feehan. First archbishop of Chicago
1830	James Knott. Governor of Kentucky
1833	Capt. John B. Pence
1857	John Worth Edmunds. Painter
1859	Katharine Bates. Writer
1862	Maurice Maderlinck. Poet
1876	Lucien Muratori. Tenor
1891	Marquis James. Writer
1898	Preston Sturges. Director
1912	Barry Sullivan. Actor
1920	Charlie Parker. Saxophonist
1938	Elliott Gould

August 30

1334	Peter I. King of Castile
1538	Ven. Cesare Baronius

1559	Cosmo Alamanni
1740	Col. Mathew Driver. Member of convention which ratified the Constitution of the United States
1748	Jacques David. Painter
1761	Ephraim Capron. Patriot
1784	Robert Walsh. Author; editor founder of the *National Gazette and Literacy Register*
1794	Stephen Kearney. "Father of the U.S. Cavalry"
1797	Mary Shelley. Author
1818	Alexander Hamilton Rice. Governor of Massachusetts
1826	Benjamin Fichel. Painter
1844	Friedrich Ratzel. Geographer
1851	William Brownell. Writer
1863	Edwin Jordan. Congressman
1883	Henry Schricker. Governor of Indiana
1893	Adm. Bernard Clarey. Director of Navy Program Planning
1896	Raymond Massey. Actor
1901	John Gunther. Author
1907	Eugene Keogh of Congress
1909	Blondell
1819	Kitty Wells
1922	Jack Ludwig. Winner of O'Henry award
1931	John Swigart. Astronaut
1939	Elizabeth Ashley. Actress
1943	Jean Claud Killy. Noted skier
1951	Timothy Bottoms

August 31

12	Caligula. Roman emperor
1569	Jahangir. Mogul emperor of India
1605	Nicolas Talon. Historian
1660	Ephraim Kidder
1792	Wilbur Fiske of the Colonies
1827	H.E. Rhoads. Wagonmaker
1830	Charles Nordhoff. Writer
1851	Samuel Van Wyck Fleet of Oyster Bay
1864	Max Zach. Musician
1879	Yoshihito. 123rd emperor of Japan
1880	Queen Wilhelmina of the Netherlands
1897	Frederick March. Actor

1898	Perry Askam. Actor
1901	William Saroyan. Author
1903	Arthur Godfrey
1906	Raymond Sommer. Race car driver
1913	Bernard Lovell. Astronomer
1914	Richard Basehart. Actor
1918	Alan Lerner. Co-author of *My Fair Lady*
1928	James Coburn. Actor
1935	Frank Robinson. Outfielder; winner of Triple Crown
1941	David Agard. Conductor
1945	Itzhak Pearlman. Pianist
1955	Edwin Moses. Gold Medal winner

September 1

866	Leo VI. Byzantium emperor
1747	Hezekia Doane. Patriot
1749	Lorenz Haschka. Poet
1770	Thomas O'Conor. Journalist
1785	Peter Cartwright
1785	Philip Allen. Formed the Rhode Island Carbineers
1792	Chester Harding. Painter
1795	James Gordon Bennett. Editor
1798	Richard Delafield
1812	James Campbell. Postmaster General
1822	Hiram Revels. Senator; chaplain of Union Army
1827	Elisha Baxter. Governor of Arkansas
1840	Mary Elizabeth Blake. Poet
1840	Joseph Tisset. Author
1849	Anastasia Ryan
1854	Engelbert Humperdinck. Composer
1858	Arthur O'Neill. Author
1860	Cleofonte Campanini. Conductor
1866	Thomas Woodlock. Editor; writer
1875	Edgar Burroughs. Writer; creator of *Tarzan*
1877	Rex Beach. Novelist
1892	Leverett Saltonstall. Senator; governor of Massachusetts
1898	Richard Arlen. Actor
1901	Adolph Rupp. Basketball coach of renown
1902	Dirk Browver. Astronomer
1906	Victoria Holt. Writer

1916   Ed Boch. National Football Foundation Hall of Fame
1922   Vittorio Gassman. Actor
1923   Rocky Marciano. Knocked out Jersey Joe Wolcott
1931   Seiji Ozawa. Conductor
1939   Rico Corty of baseball

September 2

1243   Sir Gilbert Hampshire
1251   Bl. Francis of Fabriano
1613   Etienne Champs. Author
1661   Mary Burbean of Woburn
1781   John Reed. Congressman
1783   William Jackson. A founder of the Liberty Party
1804   Pierre Ram
1824   Emeline Kimball
1834   Lucretia Hale. Writer
1838   Lilino Kalani. Last queen of Hawaii
1839   Henry George. Editor
1840   Giovanni Verga. Novelist
1848   Roxanne Boyd
1850   Eugene Field. Columnist
1859   Margaret Boyle
1866   Hiram Johnson. Governor
1870   Richard Henry Tierney. Editor
1870   Marie Ault. Actress
1877   Frederick Soddy. Author
1896   Frederick Kiesler. Architect
1911   William Harrah of the casino
1914   Ramose Bearden. Painter
1918   Allen Drury. Author
1923   Marge Champion. Dancer
1925   Ike Andrews. Congressman
1930   Kenneth Thomasma. Author
1937   Peter Ueberrath. Baseball commissioner
1938   Michael Hastings. Playwright
1946   Louis Avalon. Actor
1948   Christa McAuliffe. Astronaut

September 3

1695   Pietro Locatello. Composer
1728   Matthew Boulton. Scientist

- 1741   Owen Jones
- 1781   Fielding Lucas. Publisher
- 1797   Benjamin Webster. Actor
- 1814   Mark Hopkins
- 1820   George Hearst. Publisher
- 1821   James McClelland. Poet
- 1833   William Cushing
- 1824   Carolina Soule. Author
- 1850   Eugene Field
- 1856   Louis Henri Sullivan. A founder of the Chicago School of Architecture
- 1865   Wilhelm Bausset
- 1864   Francis Burkitt. Pioneer
- 1869   Fritz Pregl. Chemist
- 1875   Ferdinand Porsche. Automobile designer
- 1900   Sally Benson. Writer
- 1910   Dorothy Maynor. Singer
- 1917   Eddie Stanky. Baseball manager
- 1920   Marguerite Higgins. Newspaperwoman
- 1923   Ed Sprinkle of Chicago Bears
- 1931   Dick Molta. Coach of Chicago Bulls
- 1931   Bill Sadler. Race car driver
- 1935   Johnny Mathis. Singer

September 4

- 1241   Alexander III. King of Scotland
- 1757   Michael Webber. Patriot
- 1793   Edward Bates. Attorney General
- 1798   Francis Le Moyne. Abolitionist
- 1802   Marcus Whitman
- 1803   Sara Polk. First Lady
- 1805   William Dodge. Congressman; president of Houston & Texas Central
- 1806   Justin Olds. Surveyor; circuit clerk
- 1810   Donald McKay. Naval architect
- 1824   Anton Bruckner. Composer
- 1836   John Rose Hassard. Journalist; editor; author
- 1843   Patrick Kennedy. Publisher
- 1860   Hamlin Garland. Novelist
- 1867   Courtney Talbott

1905	Mary Renault. Writer
1913	Stanford Moore. Recipient of Nobel Prize
1918	Paul Harvey. News commentator
1924	Joan Aiken. Writer
1926	Ivan Illich. Author
1926	Robert Lagomarsino. Mayor of Ojai; congressman
1929	Thomas Eagleton. Senator
1931	Mitzi Gaynor. Actress
1932	Vincent Dooley. Coach
1937	Dawn Franser. Olympic swimmer
1938	Barbara Anders. Soprano

September 5

1568	Tommaso Campanella. Philosopher
1638	Louis XIV
1666	Gottfied Arnold
1672	Gottfied von Bessel. Historian
1686	Antoine Touron. Historian
1708	Isaac Seizas of the Colonies
1735	Johann Christian Bach. Composer
1791	Giacomo Mayerbeer. Pianist
1795	Étienne Taché. Statesman
1804	William Alexander Graham. Senator
1818	Charles King. Writer
1847	Jesse James
1850	Edward Butler. Editor
1867	Amy Marcy Cheney Beach. Pianist; composer
1873	David Campbell. Three time All American
1876	George Reginald Sims. "Father" of New Port Richey, Florida
1887	Rudolph Schindler. Architect
1892	Joseph Szigeti. Violinist
1901	Florence Eldridge. Actress
1912	John Cage. Composer
1924	Maria Palmer. Actress
1929	Bob Newhart
1939	Clay Regazzoni. Race car driver
1940	Racquel Welch

September 6

1475	Sebastiano Serlio. Painter
1577	Pietro Tacca. Sculptor

1697	Nicola Salvi. Architect
1702	Barent van Buren
1750	Ezra Bellows. Patriot
1766	John Dalton. Scientist; developer
1803	Commodore Jack Gillis of the Civil War
1805	Horatio Greenough. Sculptor
1831	James Gillis. Astronomer
1819	William Rosecrans. Union general; congressman; minister to Mexico
1821	Parmenas Turnley. Author
1842	Melville Ingalls of the railroad
1853	Katherine Conway. Author
1869	Catherine Aspinwall
1884	Joseph Willging. First bishop of Pueblo
1885	Otto Kruger. Actor
1890	Gen. Claire Chennault of Flying Tigers
1890	Walter Teufel of U.S. Public Health Service
1892	John Daniel Rust. Inventor of Rust Cotton Picker
1905	Christopher Weldon. Chaplain of the aircraft carrier, "Guadacanal"; bishop of Springfield, Massachuesetts
1912	Vince Di Maggio of baseball; older brother of Joe Di Maggio
1920	Barbara Guest. Author
1926	Prince Claus of the Netherlands
1929	Margaret Ballinger. Deputy Grand Matron
1947	Jane Curtin. Actress

September 7

1395	Sir Reynold de West. Baron de la Warre
1664	Johann Eckhart. Historian
1677	Stephen Hales. Inventor
1680	Joseph Kinney of the Indian Wars
1720	Archalaus Dale. Patriot
1729	Peter Lieversee of the Colonies
1769	Cassa Paine
1778	José Sanchez. Missionary
1815	John Stuart. Explorer
1819	Thomas Hendricks. Vice President of the United States
1823	Thomas Eddy. Author
1837	Joseph Dwenger. Bishop of Fort Wayne
1860	Grandma Moses. Painter

1869	Paul Nussbaum
1869	Charles O'Neill. State Supreme Court Justice
1873	Carl Becker. Author
1881	William Finn. Musician
1897	Joseph Collins. Author
1898	Henry Robinson. Author
1900	Taylor Caldwell. Novelist
1905	Ivy Baker Priest. Treasurer of the United States
1908	Paul Brown. Coach
1909	Elia Kazan. Author; director
1930	Baudouin I. King of Belgium
1936	Buddy Holly

September 8

1157	Richard the Lionhearted. King of England; crusader; gallant knight
1296	Margaret Fitz Alan
1413	St. Catherine of Balogna
1474	Ludovico Ariosto. Poet
1515	Alphonsus Salmeron. Scholar
1640	Jerome Gonnelieu. Writer
1716	Andrea Spagni. Author
1765	Pope Gregory XVI (Bartolommeo Cappellari)
1778	Clemens Brentano. Author
1789	Robert Reid. Governor of Florida
1799	Samuel Fosdick Curtis. Abolitionist
1815	Alexander Ramsey. Minnesota territorial governor (gave Ramsey County its name); negotiated treaty with Sioux; mayor of St. Paul
1821	Henry Baxter. Commander of Civil War
1827	Emil Naumann. Composer
1828	Joshua Laurence Chamberlain. Awarded Congressional Medal of Honor
1830	Frederic Mistral. Poet
1833	William Byrne
1837	Cincinnatus "Joaquin" Miller. Writer
1841	Anton Dvorak. Composer
1863	Jessie Smith. Painter
1869	Bryson Burroughs. Artist
1873	Alfred Jarry. Poet

1873	David McKay
1889	Senator Robert Taft. "Mr. Republican"
1896	Howard Dietz. Writer
1907	Robert McCaig. Writer
1908	John Cavanagh. Editor
1922	Sid Caesar
1925	Peter Sellers
1932	Patsy Cline
1960	Stephano Casiraghi

September 9

384	Flavius Honorius. Roman Emperor
1415	Frederick III. Holy Roman Emperor
1582	Penelope West of Hastings, Sussex and Boston
1629	Adm. Cornelius Tronys. As head of Dutch-Danish fleet, defeated Swedish forces
1711	Thomas Hutchinson. Governor of Massachusetts; author of *The History of the Colony of Massachusetts Bay*
1721	Judge Edmund Pendleton. A founder of the Republic; delegate to Continental Congress; drew up resolution which led to the Declaration of Independence
1731	Francisco Clavigero. Historian
1739	Luigi Galvanco
1747	Joseph Lovett of the Colonies
1750	Matthew Dance. Patriot
1765	Manuel Bocage. Poet
1767	Enos Nettleton
1770	Rachel Farnsworth
1793	Ezra McIntire
1798	Cosmo Innes
1816	Robert Tyler. Editor
1828	Leo Tolstoi. Novelist
1835	Marilda Trimmer
1863	Frederick Brand
1887	Alf Landon. Governor of Kansas; presidential candidate
1889	Glenn Dresback. Poet
1892	Tsuru Aoki. Actor
1898	Henry Styles Bridges. Senator
1899	"Schoolboy" Hoyt of baseball
1900	James Hilton. Novelist

1908	Cesare Pavese. Poet
1925	Sona Cervena. Opera singer
1934	Nedda Casei. Mezzo-soprano
1921	Otis Redding
1951	Michael Keaton. Actor
1952	Angela Cartwright. Actress

September 10

1487	Pope Julius III (Giovanni Maria Ciocchi del Monte)
1603	Henry Valois
1714	Niccola Jammelli. Composer
1754	Asa Nourse of the Colonies
1750	Nicholas Biddle. Served in Royal Navy and American Navy; commander of Andrea Doria
1791	Giusseppi Belli. Poet
1823	James Gilmore. Author
1836	Joseph Wheeler. Confederate officer; congressman; major general in Spanish-American War
1866	Jeppe Aakjaer. Novelist
1885	Berthold Altaner. Historian
1886	Hilda Doolittle. Poet
1896	Elsa Schiaparelli. Fashion designer
1898	Adele Astair. Actress
1898	Sergei Mikailoff. Pianist
1902	Jim Crowley of football
1905	Urban Nagle. Author
1905	Alfred Goldschmidt. Prominent banker
1905	Sara Martin Mayfield. Editor
1906	Leonard Lyons. Columnist
1915	Edmond O'Brien. Actor
1924	Ted Kluzewski of baseball
1929	Arnold Palmer of golf
1934	Roger Maris. Baseball immortal
1936	Tammy Overstreet

September 11

1522	Ulisse Aldrovandi
1573	Elizabeth West of Early England

1711	Alexandre Pingre of the Colonies
1786	Frederich Kublau. Composer
1790	Abaigail Davis
1792	John Goyle. Statesman
1796	Danield Barnard of Congress
1799	Lydia Steele. Pioneer
1806	Juliette Kinzie. Novelist
1821	Erastus Beadle. Publisher; issued first dime novel
1847	Mary Whitney. Astronomer
1853	Flora Stevens. Pioneer
1862	Arthur Drossaerts. First archbishop of San Antonio
1862	O. Henry
1871	George Salmon
1885	D.H. Lawrence. Writer
1890	Carl Glick. Actor; director
1891	William Thomas Walsh. Author of many works including *Out of the Wilderness, Isabella, the Last Crusader, Philip II*
1898	Antoinette Scudder. Poet
1901	D.W. Brooks. Farm organization leader
1913	Bear Bryant. Football coach
1917	Ferdinand Marcos
1919	Reed Whittemore. Poet
1924	Tom Landry. Football coach
1945	Franz Beckenbauer. Soccer player

September 12

1506	Franciscus Sonnius. Writer
1649	Bl. Giuseppe Maria Tommasi
1690	Peter Dens. Writer
1723	Johann Basedow
1747	Andrew Love. Patriot
1788	Alexander Campbell. A founder of the Disciplines of Christ
1792	Thomas Jefferson Randall. Author
1806	Andrew Foote. Civil War leader
1812	Richard Hoe. Inventor of rotary printing press
1818	Richard Gatling. Inventor
1829	Anselm Feuerbach. Painter
1829	Zenaid Fleuriot
1829	Charles Warner. Novelist
1830	James Martin Axtell

1833	Anthony Keiley. Mayor of Richmond
1850	Johann Heinrich Beck. Violinist; composer
1855	Fiona McCleod. Poet
1855	William Sharp. Author
1865	Sophus Cloussen. Poet
1880	H.L. Mencken
1881	Gerald Ames
1891	Arthur Salsberger. Publisher of *New York Times*
1892	Alfred Knopf. Publisher
1894	Daniel Feeney. Bishop of Portland
1897	Irene Curie
1898	Muriel Alexander. Actress
1901	Eileen Beldon. Actress
1912	John Merlo. Senator
1913	Jesse Owens
1931	George Jones of Country music
1958	Wilfredo Benitez. Winner of World Boxing Association junior-welterweight title

September 13

1157	Alexander Neckham. Poet
1737	Richard Bache. Postmaster General
1742	Sam Ranger. Patriot
1761	Benjamin Russel. Journalist; statesman
1786	John Rakewade. Antiquarian
1817	John McAuley Palmer
1817	Louis-Armand Garreau. Novelist
1819	Clara Schumann. Pianist
1827	Emily Haven. Author
1851	Walter Reed for whom Walter Reed Hospital is named
1854	Whillie Beard
1856	Harry Randolph Tammany
1857	Milton Hershey. Manufacturer of Hershey Bars
1859	Hon. Charles Dean Kimball. Governor of Rhode Island
1860	Gen. John Pershing
1865	Maud Booth
1868	Otokar Brenzina. Poet
1874	Henry Fountain Ashurst. Cowboy; orator; senator
1874	Arnold Schoenberg. Composer
1876	Sherwood Anderson. Writer

- 1905   Gretta Palmer. Author
- 1907   Claudette Colbert. Actress
- 1912   Horace Babcock. Astronomer
- 1924   Lauren Bacall. Actress
- 1935   Angers Morrison. Mayor of Fremont
- 1939   Arleen Aiuger. Soprano
- 1939   Richard Kiel. Actor
- 1946   Jacqueline Bisset. Actress

September 14

- 1246   John FitzAlan. Earl of Arundel
- 1246   Isabella Mortimer of England
- 1295   Sir John de Dinham of Nutwell
- 1388   Claudio Clausson. Cartographer
- 1486   Agrippa of Nettesheim
- 1543   Claudius Acquaviva
- 1618   Peter Lely. Painter
- 1643   Joseph Jouvancy. Poet
- 1717   Freelove Kilton of the Colonies
- 1739   Johann Haydn. Conductor
- 1740   Abraham Lighthill. Patriot
- 1742   James Wilson. Member, Continental Congress; Signer of the Declaration of Independence; Supreme Court Justice
- 1757   Charles Adams of the Revolutionary War
- 1758   Smallwood Acton. Patriot
- 1759   Asa Camp of the Colonies
- 1760   Maria Luigi Cherubini. Composer
- 1761   Solomon Chittenden
- 1769   Alexander Humboldt. Scientist
- 1775   John Henry Hobart
- 1792   Seba Smith. Journalist
- 1810   John Greiner. Journalist
- 1827   William Davis Bishop. Railroader, Senator
- 1830   Emily Briggs. Orator; journalist
- 1867   Charles Gibson. Originator of the Gibson Girl
- 1868   Ellen Yaw. Singer
- 1880   Davis Baker Keniston. Legislator; Chief Justice of Boston
- 1885   Vittorio Gui. Conductor
- 1892   Ben Moreel. Founder of the "Seebees"
- 1896   John Robert Powers

1898	Hal Wallis
1899	James Bellok. Novelist
1907	Edel Quinn. Missionary to Africa
1907	John Wexley. Playwright
1909	Robert Emmet Tracy. First bishop of Baton Rouge
1932	Clint Lauderdale. Ambassador to Guyana
1933	Zoe Caldwell. Actress

September 15

1613	François de La Rochefoucauld. Writer
1670	Captain Thomas Curtis. Early sea captain and delegate to the General Court
1744	Eliphalet Tisdale. Pioneer
1752	Richard Lord
1753	Ebenezer Colfax. Patriot
1765	Manuel Borage. Poet
1789	James Fenimore Cooper. Author of *Leather-Stocking Tales, The Last of the Mohicans, The Deerslayer*
1813	David Bacon. First bishop of Portland, Maine
1818	Justus Roth. Mineralogist
1824	Joseph Hergenröther. Historian
1834	Heinrich Treitschke. Historian
1839	Broxton Cash
1857	William Howard Taft. President of the United States and Chief Justice of the Supreme Court
1858	Charles de Foucauld. Playboy to hermit; wrote *Rule for the Little Brothers of the Sacred Heart*
1881	Ettore Bugatti. Race car designer
1884	Jean Renoir. Writer; director
1888	Antonio Assari. Winner of Italian Grand Prix
1889	Robert Benchley
1890	Frank Martin. Composer
1890	Claude McKay. Poet
1890	Agatha Christie. Author
1894	Jean Renoir
1903	Ray Acuff. Bandleader
1903	Roy Claxton of "The Smoky Mountain Boys"
1904	Humbert II. King of Italy
1905	Jean Tennyson. Opera singer
1917	Hilda Gulden. Soprano

1918	Myron Goldsmith. Architect
1922	Jackie Cooper. Actor
1934	"Fob" James. Governor of Alabama
1938	Gaylord Perry. Pitcher
1946	Tommy Lee Jones. Actor
1961	Dan Marino of football

September 16

1386	St. Ambrose of Camaldoli
1625	St. Gregory Barbarigo
1654	Philippe Avril. Early traveler and chronicler
1713	Gideon Watrous
1730	Franz Erthol of the Colonies
1733	Abraham Whipple. Leader in the Revolutionary War
1756	David Currier. Patriot
1797	Levi Ives
1798	William Goode. Legislator
1803	Orestes Brownson. Writer; editor; established *Boston Quarterly Review*, wrote *The Infidel Converted, The American Republic*
1816	Charles Newton. Archeologist
1823	Francis Parkman. Historian
1831	Henri Cosgrain. Author
1836	Edward Funston. Congressman
1838	James Hill. Railroad builder
1842	Charles Fosdick. Author
1850	Robert Barr. Novelist
1853	Clemens Beaumker. Historian
1868	Edmund Gibbons. Co-founder of the National Legion of Decency
1875	J.C. Penney. Founder, chain stores
1886	Alfred Noyes. Poet
1893	Sir Alexander Korda. Producer
1923	Janis Paige. Actress
1927	Peter Falk. Actor
1927	Jack Kelly. Mayor
1932	Anne Francis. Actress
1950	David Bellamy of the Bellamy Brothers
1958	Orel Hershiser

September 17

879	Charles. King of France
1312	William de Burgh. Fourth Earl of Water

1550	Pope Paul V (Camillo Borghese)
1722	Samuel Adams
1730	Frederick William Von Stecchen. Served as captain under Frederick the Great in Seven Years War; served under George Washington at Valley Forge; Inspector General of Continental Army; fought at Battle of Monmouth
1746	Benjamin Longrath of the Colonies
1758	David White. Patriot
1785	Diego Garcia. First bishop of California
1880	Franklin Buchanan. Commodore of the "Merrimack"
1815	Arthur St. Leon of ballet
1819	Marthinus Pretorius. First president of South African Republic
1825	Lucius Lamar. Senator; Secretary of the Interior
1826	Georg Riemann. Mathematician
1834	Thankful Wells. Pioneer
1836	Gen. William Jackson Palmer. Founder of Colorado Springs
1843	William Elliot Griffis
1854	Effie Ellsler. Actress
1899	Earl Webb of New York Giants, Chicago Cubs, Boston Red Sox, Detroit Tigers and Chicago White Sox
1900	J. Willard Marriott
1908	John Creasey. Writer
1916	Mary Stewart. Author
1917	Louis Auchincloss. Writer
1922	Vance Bourjaily. Novelist
1923	Hank Williams. Country singer
1929	Stirling Moss. Race car driver
1930	Edgar Mitchell. Astronaut
1931	Anne Bancroft. Actress
1936	Ken Kesey. Writer
1937	Orlando Cepeda
1938	Lee Roy Yarbrough. Winner of Dixie 500, of the Daytona 500, the Rebel 400 and the World 600

September 18

53	Trojan. Roman emperor
1596	James Shirley. Poet
1709	Samuel Johnson. Noted author
1729	Jabez Beebe. Patriot
1733	George Reed. Delaware signer of the Declaration of Independence; member of Continental Congress

- 1798 Rufus Babcock
- 1799 John Kingsley
- 1812 Herschel Vespasian Johnson. Vice presidential running mate of Stephen A. Douglas
- 1818 Mary Ann Vincent. Actress
- 1827 John Townsend Trowbridge. Author
- 1838 Anton Mauve. Painter
- 1838 Frank Shapleigh
- 1848 Henry Smith. Pioneer; traveled overland by ox-team
- 1861 Owen Seaman. Poet
- 1870 Clark Wissler. Anthropologist
- 1878 James Richardson
- 1893 Arthur Benjamin. Composer
- 1894 Foy Compton. Actress
- 1895 John Diefenbaker. Prime minister of Canada
- 1905 Greta Garbo
- 1905 Eddie "Rochester" Anderson
- 1910 Francis Vanderbilt. Writer
- 1956 Debbie Fields. Created "Mrs. Field's Cookies"
- 1968 Tony Kukoc of Chicago Bulls
- 1971 Lance Armstrong. Champion Cycler

September 19

- 1736 Henry Clay. Patriot
- 1737 Charles Carroll of Carrollton. Signer of the Declaration of Independence
- 1749 Jean Baptiste Joseph Delambre. Astronomer
- 1764 Abraham van Winkle
- 1794 Charles Boyd. Established state route from Peoria to Galena
- 1799 Azuba Ricker. Pioneer
- 1814 Karl Savigny. Statesman
- 1821 Gilbert Haven. Teacher; chaplain; editor
- 1832 Moses Armstrong. Congressman
- 1847 Tomas Camara y Castro. Bishop of Salamania
- 1854 Thomas Meehan. Editor; historian
- 1855 Katharina Klafsky. Singer
- 1855 Ernest Acheson. Congressman
- 1862 Michael Gatherer
- 1865 Charles Benza. President of Thiel College; author
- 1866 James Barnes. Writer

1869	William Amory. Industrialist
1879	Gen. Hugh Drum. Inspector General; Head of Eastern Defense Command; President of the Empire State Building corporation
1894	James W. Alexander. Mathematician
1895	Francis Cotton. First bishop of Owensboro
1907	Lewis Powell, Jr. Supreme Court Justice
1911	William Galding. Novelist
1922	Emil Zatopeh. Long distance runner
1926	Marian La Follette. Legislator
1930	Michal Abrams. Pianist; composer

September 20

1501	Gerolamo Cardano. Mathematician
1667	Charles Hugh of the Colonies
1744	Giacomo Quarenghi. Architect
1751	Nathan Ranney. Patriot
1762	Pierre Fontaine. Painter
1775	Herman Oviatt of Goshen
1797	Philipp Reiss. A founder of the town of Guttenberg, Iowa, on the Mississippi River
1799	George Hallms. Commander of Naval Station at Richmond, Charlotte and Wilmington during Civil War
1807	Charles Dolman. Publisher
1809	Sterling Price. Confederate general; legislator; governor
1822	Henry James Cooleridge. Writer
1838	Vincent Borzynski. Missionary
1838	John Ming. Author
1863	William Grimm. Co-author of Grimm Fairy Tales
1865	Edward Brewster
1870	Thomas Wayland Vaughn. Oceanographer
1878	Upton Sinclair. Novelist
1880	Ildebrando Pizetti. Composer
1885	Jelly Roll Morton. Jazz musician
1886	Sister Kenny. Pioneer in treatment of polio
1898	Chuck Dressen. Baseball manager
1917	Red Auerbach. Renowned basketball coach
1930	Walter Roorda. Legislator
1934	Sophia Loren. Actress
1951	Guy Lafleur of ice hockey

September 21

- 1415 Frederich III. Holy Roman Emperor crowned in 1452
- 1629 Philip Howard
- 1645 Louis Joliet. Explorer
- 1668 Roman Hinderer. Missionary
- 1751 Alexander Vining
- 1758 Christopher Gore. Senator
- 1804 Simeon Nash. Author; senator
- 1809 Sophie Hawthorne. Artist
- 1817 Joseph Finatti. Missionary; literary editor of *Boston Globe*
- 1827 Michael Corcoran. Commander of the Irish Leagion in American Civil War; served at Battle of Bull Run
- 1827 John Massilon McConike
- 1833 Augustus Myers
- 1842 Abdal-Hamid. Last Ottoman sultan
- 1842 Roy Zearing
- 1861 Daniel Quinn. Author
- 1861 Edson Shiffle
- 1866 H.G. Wells. Author
- 1870 Frank Butterworth. Fullback
- 1874 Gustave Holst. Composer
- 1876 Herman Bernstein. Author
- 1882 Mary Becker
- 1931 Larry Hagman. Actor; appeared in "The Edge of Night," "The Eagle Has Landed"
- 1933 Dick Simon. Race car driver
- 1947 Stephen King. Novelist
- 1950 Bill Murray. Actor

September 22

- 1577 Christoph Besold. Writer
- 1601 Anne of Austria. Regent
- 1636 Sarah Stearns. Pioneer
- 1644 Jacques Echard. Historian
- 1665 Cpl. Richard Kimball of Ipswich
- 1728 Francis Cutting. Patriot
- 1756 John Dubbs
- 1787 Pasquele Gizzi
- 1788 Theodore Hook

1790	Commodore Cornelius Kinchiloe-Stribling. Ship captain of the Antebellum period
1827	Peter Turney. Governor of Tennessee
1828	Theodore Winthrop. Poet
1829	William Belknap. Secretary of War
1845	Hartmann Grisar. Historian
1855	John Talbot Smith. Author
1862	Maurice Barres. Novelist
1872	Eleanor Abbott. Author
1872	Walter Rothwell. Conductor of Los Angeles Philharmonic
1877	Victor Shelford
1878	Shigeru Yoshida. Ambassador; statesman
1879	John La Gorce. Geographer; writer; postmaster of Little America, Antactica
1885	Erich von Stroheim. Actor
1892	Frank Sullivan. Writer
1895	Babette Deutsch. Poet
1895	Herbert Johnson. Baritone
1895	Paul Muni. Actor
1902	John Houseman. Actor
1903	Howard Jarvis. Tax reformer
1956	Debbie Boone. Singer

September 23

63 BC	Caesar Augustus
1003	Alan Villiers. Author of *Son of Sinbad*, *The Coral Sea*
1439	Francesco Giorgio. Painter
1602	Alice Baynton of Boston
1647	Governor Joseph Dudley
1678	Gideon Chittenden
1721	Hans Hagey
1733	David Gale
1745	John Sevier. First governor of Tennessee
1747	Elizabeth Bradford
1752	Samuel Longgen. Patriot
1782	Prince Maximilian
1783	Peter Cornelius. Artist
1786	John England. Bishop of Charleston; editor; writer
1800	William McGuffey. Editor of McGuffey readers which promoted moral education

1820	Gen. Thomas Kilby Smith of the Civil War
1823	Sara Lippincott. Writer
1833	Everett Ricker. Hero of Civil War
1838	Victoria Claflin Woodhull
1852	James Beckwith. Painter; promoter of Art Guild
1861	Robert Bosch. Inventor of Sparkplugs
1865	Suzanne Valadon. Painter
1871	Frantisek Kupka. Abstract painter
1884	Eugene Talmadge. Governor of Georgia
1889	Walter Lippman. Journalist
1902	Frederick Piper. Actor
1911	Adm. Reuben Whitaker
1930	Ray Charles. Singer; pianist
1938	Romy Schneider

September 24

1301	Sir Ralph. First Earl of Stafford
1583	Albrecht Wallenstein. Military leader of Thirty Years War
1755	John Marshall. Chief Justice of the Supreme Court
1796	Antoine Barye. Sculptor
1798	Louisa Clap
1811	Alden Booth
1815	George Stafford
1816	Samuel Fifield
1823	Fredrick Maassen. Scholar
1825	Frances Harper. Poet
1834	Joseph Riley
1837	Marcus Hanna. Senator
1851	George Hoffman
1855	Daniel Feehan. Bishop of Fall River
1857	Sir Ben Greet. Shakespearean actor
1858	Hugh Kelly. President of Maritime Exchange; member of New York State Commerce Commission
1867	Margaret Neilson Armstrong. Author; illustrator
1891	Karin Bronzell. Mezzo-soprano
1896	F. Scott Fitzgerald. Novelist
1900	Claire Adams. Actress
1916	Hollis Alpert. Writer
1920	Mickey Rooney
1921	Jim McKay. Sports commentator

1936 James Henson. Puppeteer
1947 Roxy Allessandro. Writer

September 25

1358 Yoshimitsu Ashi Kaga. Shogun of Japan
1599 Francesco Barronimi. Architect
1728 Mary Otis Warren. Author
1740 Joseph Ruffner
1740 Adam Vrooman
1793 John Neal. Novelist
1806 James Ward. Commander of the "Cumberland"; first Union Naval officer to die in Civil War
1829 William Rossetti. Writer
1838 Adm. Joseph Trilley of Battle of Mobile Bay
1841 Nicholas I. King of Montenegro
1855 Adm. William Benson. Chief of Naval operations; chairman of United States Shipping Board; first president of National Council of Catholic Men
1858 Arthur Hecker. Painter
1866 Thomas Hunt Morgan. Nobel Prize recipient
1883 Arthur Byne. Author
1897 William Faulkner. Novelist
1902 Herbert Ashton. Actor
1905 "Red" Smith. Sportswriter
1917 Mary Clabaugh Wright. Historian
1925 Silvana Pampanini. Actress
1926 Aldo Ray
1927 Sir Colin Davis. Conductor
1929 Barbara Walters
1942 Henri Pescaralo. Race car driver
1945 Michael Douglas
1946 Felicity Kendall. Shakespearean actress
1951 Mark Hamill. Actor
1952 Christopher Reeve. Actor

September 26

1651 Francis Pastorius
1706 Samuel Webster

1749	Jacob Cushing. Patriot
1762	George Christian of the Revolutionary War
1791	Theodore Gericault. Artist
1796	Richard Henry Bayard. Senator
1809	Philipp Jolly. Scientist
1821	Thomas Harper
1841	Thomas Beach. Author of *Twenty-five Years in the Secret Service*
1859	Irving Bacheller. Novelist
1862	Arthur Davis. Painter
1866	Montgomery Heston
1888	Terence Connolly. Author
1888	Frank Dabie. Folklorist
1888	Thomas Eliot. Poet
1891	Charles Munch. Conductor of Boston Symphony
1897	Pope Paul VI (Giovanni Montini)
1898	George Gershwin. Composer
1917	Joe Tucker of Cleveland Indians
1925	Marty Robbins
1944	Sam Posey. Race car driver
1947	Lynn Anderson
1948	Olivia Newton John
1962	Melissa Anderson. Actress

September 27

1389	Casimo de Medici of Florence
1601	Louis XIII. King of France
1627	Jaques Bassuet. Orator
1643	Solomon Stoddard. Puritan leader
1648	Michelangelo Tamburini
1694	Eleazer Allen. Great-grandson of Priscilla and John Alden
1722	Samuel Adams. Signer of the Declaration of Independence; organizer of the Sons of Liberty and the Boston Tea Party
1772	Martha Jefferson Randolph
1783	Augustín de Iturbide. Emperor of Mexico
1789	René Rohrbacher. Historian
1793	Denis Alfre. Archbishop of Paris
1796	Henry Lemcke. Missionary
1803	Samuel Francis Du Pont of the Navy
1808	James Madison Trimble. President of Hillsboro & Cincinnati Railroad

1809	Adm. Raphael Semmes. Commander of the "Sumpter"; author of *Confederate Raiders*
1810	Cammillus Tarquini. Archiologist
1817	Paul Fenal. Novelist
1823	William Goodwin. Author
1827	Charles Boyd Curtis. Author; patron of Metropolitan Museum of Art and American Museum of Natural History; member of Union League
1840	Alfred Mahan. Naval officer; historian
1840	Thomas Nast. Cartoonist
1857	Adm. Robert Griffin. Chief, Bureau of Steam Engineering; editor; writer
1871	Grazia Deledda. Winner of Nobel Prize
1871	Martin Glynn. Governor of New York
1874	Myrtle Reed. Author of *Lavender & Old Lace*
1900	Clementine Paddleford. Writer
1907	Abraham Duker. Author
1916	Martha Scott. Actress
1922	Arthur Penn. Director
1944	Arthur McCue
1949	Mike Schmidt of baseball
1958	Shawn Cassidy. Actor

September 28

1639	Lord Russell. Statesman
1698	Pierre Moupertuis. First to measure curvature of the earth
1741	Jacob Cuyler. Patriot
1745	Asa Gage of Revolutionary War
1747	Job Shippey of the Colonies
1763	James Burson. Patriot
1768	Christian Schoettler
1787	Jean Gasselim. Author
1794	Simion Ides. Painter
1803	Prosper Mérimée. Writer
1839	Frances Willard
1852	Henri Moissan. Discoverer of fluorine
1856	Kate Wiggin. Writer
1859	Kale Riggs. Author
1872	Lena Ashwell. Actress
1883	James Macelwane. Scientist; president of Seismological Society; author

1896	Elizabeth Lynskey. Writer; lecturer
1898	Booke Carter. Broadcaster
1902	Ed Sullivan
1909	Al Capp. Cartoonist; creator of "Lil Abner"
1916	Peter Finch. Actor
1924	Marcello Mastrovanni
1931	Barbara Little of Lancaster City Council
1934	Brigitte Bardot
1937	Edward Applebaum. Composer
1939	Ross Johnson. Legislator

September 29

106BC	Pompey the Great
1402	Bl. Ferdinand. Infante of Portugal
1547	Miguel Cervantes. Poet
1598	John Emery. Landed at Boston on the ship "James" in 1635
1610	Gabriel Drevillets. Missionary
1631	Jean Pierron. Missionary
1640	Charles Coysevox. Sculptor
1691	Bishop Richard Challoner. Author
1735	Theobold Schott
1755	George Farragut. Navy hero of Savannah and Charleston
1758	Lord Horatio Nelson
1803	Adrian Richter. Painter
1819	John Luers of Oldenburg
1820	Bernard O'Reilly. Scientist
1859	Charles Townsend. Director of New York Aquarium
1895	Roscoe Turner. Racer; stunt flyer
1901	Enrico Fermi
1907	Gene Autrey. Singer
1908	Greer Garson. Actress
1913	Stanley Kramer. Producer
1916	Trevor Howard. Actor
1927	Paul McCloskey. Statesman
1931	Anita Ekberg. Actress
1935	Jerry Lee Lewis

September 30

1633	Elizabeth Aspinwell
1663	Isaac Chittenden

1715	Étienne Condillac
1719	Thomas Hazen. Pioneer
1737	Sarah Packard Snell
1751	Elhanan Winchester
1773	Ichabod Eastman
1791	William Peck Houghton
1794	Karl Begas. Painter
1802	Antoine Balard. Discovered bromine
1821	Robert Hay of Milo
1823	Ferdinand Sutton
1833	Matthew Quay. Senator
1835	John Page
1859	Julius Lichtenbert
1862	William Lockeye. Actor
1863	Peter Fuchs
1870	Jean Perrin. Nobel Prize winner
1876	Harry Cosey
1877	John Perry
1878	Albert Gille. Journalist
1898	Renee La Fonte. Actress
1898	Princess Charlotte. Duchess of Valentinois
1909	Patrick Skehan. Scholar
1924	Truman Capote
1927	William Mervin. Poet
1941	Reine Wisell. Racer
1931	Angie Dickinson
1935	Johnny Mathis
1941	Sigrid Abbott. Journalist

October 1

1207	Henry III. King of England
1507	Giacomo Vignola. Architect
1730	Richard Stockton. Signer of the Declaration of Independence who signed away his health and life when he affixed his name to the document
1748	Benjamin Goodhue. Senator
1820	Nathaniel Boyd
1832	Caroline Harrison. First Lady; artist; first president of the Daughters of the American Revolution
1858	Frank von der Stucker. Musician

- 1860 Amos Butler. Zoologist
- 1865 Paul Dukas. Composer
- 1881 William Boeing. Aircraft developer
- 1885 Louis Untermeyer. Editor; author; lecturer
- 1893 Faith Baldwin. Novelist
- 1894 Ralph McKenzie. Coach; inducted into Eureka College Hall of Fame, Northern Illinois Hall of Fame and Greater Peoria Hall of Fame; recipient of Timmie Award and Distinguished Service and Coaching Award from McGregor Sports Foundation
- 1899 Ernest Haycox. Writer
- 1904 Vladimir Horowitz. Pianist
- 1910 Bonnie Parker of Bonnie and Clyde
- 1911 Herman Hickman. All American
- 1920 Walter Mattheau. Actor
- 1921 William Rehnquist. Supreme Court Justice
- 1924 James Earl Carter, Jr. President of the United States
- 1927 Tom Bosley. Actor
- 1928 George Peppard. Actor
- 1932 Frank Gardner. Race car driver
- 1933 Bonnie Owens. Singer
- 1935 Julie Andrews. Singer
- 1936 Edward Villella. Dancer
- 1945 Rod Carew of Minnesota Twins and California Angels

October 2

- 1452 Richard III. Last English King of the House of York
- 1538 St. Charles Borromeo
- 1687 Daniello Concina
- 1737 Francis Hopkins. New Jersey signer of the Declaration of Independence
- 1741 Augustin Barrnel. Patriot
- 1754 Samuel Shaw. Revolutionary War Officer
- 1755 Hannah Adams. Historian
- 1756 Daniel Barber of Continental Army. Author of *History of Our Times*
- 1761 Charles Nerinckx. Missionary
- 1763 Ebenezer Burgess
- 1798 John Lewis Gervois of Continental Congress
- 1798 Mother Theodore Guirin

1831	Edwin Godkin. Journalist
1847	Paul von Hindenburg
1854	Leonard Ochtman. Painter
1875	Michael Earls. Writer
1877	Carl Hayden. First congressman elected from Arizona, he served fifteen years, followed by forty-two years in the Senate — the longest stay in history
1886	Robert Trumpler. Astronomer
1888	DeWitt Clinton Ramsey. World War II Naval Aviation leader
1894	Adm. Thomas Sprague
1895	Groucho Marx
1896	Bud Abbott of Abbott & Costello
1946	Mark Baker. Actor
1957	Gail Pennington of Broadway
1962	Alexander de Jong. Actor

October 3

1357	Sir Robert de Ferrers of Chartley
1390	Humphrey. Duke of Gloucester
1458	St. Casimir
1554	Fulke Greville. Poet
1716	Giovanni Beccarea. Physicist
1761	William Lester
1790	Coowescoowe. Tribal leader; president of Cherokee Council; Chief of United Cherokee Nation
1799	Gen. Lucas Pond of Pondsville. Senator
1800	George Bancroft. Historian
1804	Townsend Harris
1858	Eleonora Duse. Actress
1857	Gertrude Atherton. Novelist
1867	Pierre Bonnard. Painter
1870	Felix von Kraus. Opera singer
1873	Emily Post
1876	George Barrere. Flutist
1882	Alexander Jackson. Artist
1894	Elmer Robinson. Mayor of San Francisco
1886	Barbara Karinska. Academy Award winner
1893	Claud Allister
1899	Gertrude Berg. "Molly Goldberg" of radio
1900	Thomas Wolfe. Writer

1925	Gore Vidal. Writer
1928	Erik Bruhn. Ballet dancer
1935	Charles Duke, Jr. Astronaut
1936	Steve Rich. Composer
1942	Judy Ross

October 4

1331	James Butler. Second Earl of Ormond
1532	Francisco de Toledo. Theologian; author of many works on philosophy
1542	St. Robert Bellarmine
1579	Guido Bentioglio. Writer
1602	François Le Mercier
1607	Francisco de Rojas. Poet
1633	James II. King of England
1708	Antonio Vezzosi. Writer
1716	James Lind. Discovered that citrus juice prevents scurvy
1720	Giambattista Piranesi. Architect
1728	Jean Cosat
1749	Jean Duport. Cellist
1761	Iriah Vermillion
1762	Casper Luhtenberger
1796	French Forrest of Civil War
1796	John Richardson. Novelist
1810	Eliza Johnson (Mrs. Andrew Johnson)
1814	Jean François Millet. Painter
1815	Franz Clemens. Philosopher
1822	Rutherford B. Hayes. President of the United States
1822	James Garretson. Author
1879	Edward Murray East. Biologist
1884	Damon Runyon. Columnist
1887	Charles Buddy. First bishop of San Diego
1893	Walter Maier. Radio preacher
1895	Buster Keaton. Actor
1897	Ralph Gorman. Editor; author
1924	Charlton Heston

October 5

1696	St. Alphonsus Maria Liguori. Founder of the Congregation of the Most Holy Redeemer and author of one hundred and eleven books

1703   Jonathan Edwards. Writer
1720   Charles Cordell
1751   James Iredell. Supreme Court Justice
1762   Gould Davenport. Patriot
1781   Bernhard Bolzano. Mathematician
1830   Chester Arthur. President of the United States
1840   Prince John II of Lichtenstein. Reigned over seventy years, one of the longest reigns in history
1845   Eugene Garvey. Bishop of Altoona
1846   Francis Gosquet
1850   William Gibson. Artist
1858   Prince Henry Maurice of Battenberg
1864   Louis Lumiere. Inventor
1873   Louis Betts. Artist
1879   John Erskine. Writer
1895   Gen. Walter Bedell Smith. Statesman; author
1898   Leo Allen of Congress
1902   Roy Kroc. Founder of McDonalds
1919   Donald Pleasence. Writer
1929   Richard Gordon, Jr. Astronaut
1934   Robert Anderson. Organist
1937   Gene Albright. Producer
1939   Marie Claire Blais. Author
1942   Charles Ansbacher. Conductor
1951   Karen Allen. Actress

October 6

1707   Thomas Falkner
1714   Giovanni Costadori. Historian
1714   Dom Anselmo. Historian
1732   Nevil Maskelyne. Producer of Nautical Almanac
1744   James McGill. Established McGill University
1753   Christian Blickensderfer. Patriot
1771   Jeremiah Morrow. First congressman from Ohio; governor
1777   Guillaume Dupuytren. Anatomist
1784   Pierre Dupin
1794   Charles Wilkins Short. Botanist
1795   Joshua Giddings. Statesman
1801   Theophilus Miles
1812   James Mulholland. Designer of locomotives

1820	Jenny Lind. Singer
1822	Fenner Kimball. Legislator
1846	George Westinghouse of Westinghouse Electric
1861	Matteo Ricco
1862	Albert Beveridge. Orator; biographer of Lincoln
1868	Christopher Ward. Historian
1882	William Cutter Hunt, "Mr. Golf"
1886	Edwin Fischer. Pianist
1887	Maria Jeritza. Opera singer
1895	Caroline Gordon. Novelist
1897	David Dietz. Writer
1903	Brien McMahon. Judge; senator
1908	Carole Lombard. Actress
1913	Richard Dyer-Bennett. Singer
1915	Humberto Cardinal Medeiras. Archbishop of Boston
1916	Stanley Ellin. Writer
1937	Fritz Scholder. Artist
1944	Carlos Pace. Race car driver
1956	Jimmy Cefalo of football

October 7

1655	Caspar Castner. Missionary
1708	Henry Maynard
1715	Michel Benoit. Astronomer
1727	William Samuel Johnson. Participated in Stamp Act Congress; first president of Columbia College
1735	Silas Chase
1747	Ebenezer Zane. Pioneer founder of Zanesburg
1754	Ebenezer Goodhue. Patriot
1819	Frederick Record. Author
1826	William Bate. Governor of Tennessee
1832	Charles Converse. Composer
1842	Kate Bateman. Actress
1842	Bronson Howard. Playwright
1849	James Riley. Poet
1858	Joseph Ransdall. Senator
1861	Dudley Phelps. Writer
1888	Henry Wallace. Vice President of the United States
1897	Myles Connelly. Author
1901	Prince Souwanne Phouma of Laos

1904 Chuck Klein of Baseball Hall of Fame
1907 Helen MacInnes. Writer
1913 Eleazer Bromberg. Mathematician
1917 June Allyson. Actress
1926 Diane Lynn
1930 Curtis Crider. Grand Prix driver
1943 Oliver North. Marine officer; awarded Silver Star, Bronze Star, Purple Heart

October 8

1553 Jacques Thou. Historian
1885 Heinrich Schwartz. Composer
1609 John Clarke. Established town of Portsmouth
1676 Benito Jeronimo. Writer
1697 William Smith. A founder of Kings' College
1740 Edmund Williams of the Colonies
1778 Hyacinthe Quelan
1794 Caroline Howard. Author
1810 James Marshall. Discovered gold in Caiifornia
1828 Isaac Gray. Governor of Indiana
1836 Guiseppina Marlacchi. Dancer
1838 John Hay. President Lincoln's Secretary of State; ambassador to Great Britain
1846 Tarleton Bean. Zoologist
1857 Edward Albee of Vaudeville
1861 Marie van Zandt. Soprano
1890 Eddie Rickenbacker. Flying Ace of World War I and World War II; race car driver
1894 Gerald Ellard. Lecturer; writer
1894 Pat Scanlan. Editor
1899 Meggie Albanesi. Actress
1901 Marcus Oliphant. Scientist
1904 George Heath. Race car driver
1919 Jack McGrath. Racing great
1925 Irene Dalis. Mezzo-Soprano
1926 Hollingsworth McMillion. Race car driver
1940 David Carradine. Actor
1943 Chevy Chase. Actor

October 9

- 1221 Ognibene. Thirteenth Century scholar
- 1261 Diniz. King of Portugal
- 1623 Ferdinand Verbiest. Missionary; astronomer
- 1663 Giovanni Crescimbeni. Historian
- 1663 Francis Schmalzgruber. Author
- 1747 Thomas Coke
- 1750 Hugh Robertson. Patriot
- 1755 Richard Furman of the Colonies
- 1782 Louis Cass. U.S. Marshall; senator; governor of Michigan Territory; Secretary of War
- 1795 Josiah Tattnall. Captain of the Confederate Navy
- 1805 William Gwin. Established San Francisco Mint
- 1829 Richard Ayer. Congressman
- 1830 Harriet Hasmer. Sculptor
- 1832 Elizabeth Allen. Poet
- 1835 Camille Saint Saens. Composer
- 1839 Winfield Scott Schley. Navy officer; author of *Forty-five Years Under the Flag*
- 1848 Frank Duveneck. Artist
- 1863 Elaine Goodale. Poet
- 1863 Gamaliel Bradford. Writer
- 1863 Edward Bok. Editor
- 1868 Mary Hunter Austin. Writer
- 1873 Carl Flesch. Violinist
- 1890 Aimee Semple McPherson. Evangelist
- 1897 John J. Considine. Author
- 1899 Bruce Cotton. Historian of Civil War
- 1908 Jaques Toti. Director
- 1909 Reuben Kramer. Sculptor
- 1919 Irmgard Seefried. Soprano
- 1925 Robert Finch. Secretary of HEW

October 10

- 1583 Hugo Gratius
- 1609 Giovanni Bona
- 1610 St. Gabriel Lalemont. North American martyr
- 1710 Alban Butler. Historian
- 1731 Henry Cavendish. Pioneer in gravitation

1738	Benjamin West. Painter
1738	Capt. Peter Boylston Adams. Brother of President John Adams
1741	Agatha Deken. First Dutch novelist
1752	Hezekiah Love. Patriot
1769	Mariano Payeras. Missionary
1777	Hezekia Niles. Editor
1788	Joseph Eichendorff. Writer
1796	Ichabod Goodwin. Governor of New Hampshire
1811	Joseph Jukes. Geologist
1813	Giuseppi Verdi. Composer
1830	Isabella II of Spain
1833	John Studebaker. Wagonmaker; producer of automobiles
1837	Robert Gould Shaw of the Civil War
1853	Victor Howard Metcalf. Secretary of Commerce; Secretary of Labor
1859	Maurice Prendergast. Painter
1860	Rufus Reading. Statesman
1889	Kermit Roosevelt
1900	Helen Hayes
1906	Paul Creston. Composer
1918	Thelonious Monk. Jazz musician
1930	Eugenio Castelotti. Race car driver
1958	Tanya Tucker

October 11

1573	Jaques Booner
1671	Frederick IV. King of Denmark
1693	Stephen Doubrelau. Missionary
1751	Abner Sackett of the Colonies
1754	Samuel Lowell. Patriot
1758	Heinrich Olbers. Astronomer
1764	Prentiss Mellen. Senator
1792	Anthony Blanc. First archbishop of New Orleans
1806	Alexander. Prince of Serbia
1813	Louis Venillot. Journalist; writer
1825	Conrad Mayer. Writer
1834	Gen. John Coppinger. Served in Civil War, Indian Wars and Spanish-American War
1836	Charles McCabe. Chaplain in Union Army
1844	Henry Heinz. Marketed pickles

1855	Amy Lee. Opera singer
1857	James Monaghan. Editor
1872	Harlan Fiske Stone. Supreme Court Justice
1876	Gertrude von Le Fort. Poet
1885	François Mauriac. Nobel Prize winner
1887	Willie Hoppe of billiards
1891	Edwin Dickinson. Painter
1897	Joseph Auslander. Poet
1910	Joseph Alsop. Journalist
1918	Jerome Robbins. Director
1928	Alfonso de Portago. Jockey; race car driver
1929	Vivian Matalon. Director
1939	Maria Bueno. Tennis star

October 12

1537	Edward VI. King of England
1565	Bl. Ippolito Galantini
1710	Jonathan Trumbull. Governor of Connecticutt
1725	James Lowell. Patriot
1731	Jean Baptiste Blanchard. Writer
1813	Lyman Trumbull. Senator
1815	William Hardee. Confederate general
1839	Sarah Randolph. Author
1839	Dom Sebastian
1844	George Washington Cable. Writer
1855	Arthur Nikisch. Conductor
1856	Thomas Ewing Sherman. Missionary
1872	Ralph Williams. Composer
1881	William Agnew. Editor; president of Creighton University
1889	Perle Mesta
1889	Dietrich von Hildebrand. Philosopher; author of more than thirty books, including *Fundamental Moral Attitudes, Christian Ethics, The Devastated Vineyard, Trojan Horse in the City of God*
1904	Elinor Rice. Author
1906	Joe Cronin of baseball
1906	Janet Gaynor. Actress
1908	Paul Engle. Poet
1911	Ann Petry. Writer
1912	Francis Fink. Editor

1926	Cesar Pelli. Architect
1932	Dick Gregory
1937	Paul Hawkins. Winner of Monaco Grand Prix
1951	Susan Anton. Actress

October 13

1700	Charles Toustain of the Colonies
1716	Baldwin Dade. Patriot
1754	Molly Pitcher (Mary Ludwig Hays McCauley). Carried water in a pitcher to the wounded soldiers at the Battle of Monmouth; took her husband's place at a cannon when he became incapacitated
1763	Jonah Scoggins
1778	William Marks. Senator
1800	René Caron. Statesman
1821	Rudolph Virchow. Scientist
1826	Johanna Wagner. Soprano
1826	La Fayette Baker
1843	Thomas Dight. Anatomist
1862	Mary Kingsley. Author
1865	Bird McGuire of Congress
1870	Della Fox. Actress
1872	Louise Hale
1882	Louis Kenedy. Publisher
1886	Bernard Bell. Author
1890	Conrad Richter. Novelist
1893	Oscar Youngdahl
1897	Irene Rich. Actress
1902	Arna Bontemps. Writer
1902	William Shaw. Race car driver
1907	Yves Allegret. Director
1910	Art Tatum of jazz fame
1915	Wesley Powell. Governor of New Hampshire
1920	Laraine Day. Actress
1931	Betty Abbott. Biologist

October 14

1425	Alesso Baldovinetti. Artist
1644	William Penn

1660	Hannah Baker. Pioneer
1696	Samuel Johnson. Philosopher
1730	David Leighton of the Colonies
1763	William Rogers
1784	King Ferdinand VII of Spain
1819	Ben Holladay. Builder of the American West
1824	Adolphe Monticelli. Painter
1834	Alphons Huber
1857	Elwood Haynes. Inventor
1871	Carl Rathberger
1875	Thomas Walsh
1879	Ernest Sutherland Bates
1882	Eamon de Valera. Prime Minister and president of Ireland
1888	Katherine Mansfield. Writer
1890	Gen. Dwight Eisenhower. President of the United States
1899	Lillian Gish. Actress
1905	J. Franklin Ewing. Anthropologist; writer
1908	Frederick Castle. Recipient of Medal of Honor
1908	Harmon Bellamy. Writer
1909	Bern Rosemeyer. Race car driver
1916	Jack Arnold. Director
1939	Ralph Lauren. Designer
1942	Lee Arthur. Actress
1947	George Maull. Conductor
1950	Sheila Young. Skater; cyclist

October 15

1331	Elizabeth de Meinell of Whorlton
1542	Akbar the Great. Mogul Emperor of India
1608	Evangelista Torricelli. Mathematician
1671	Lewis Morris. Governor of New Jersey
1701	St. Marguerite of Youville. Foundress of Sisters of Charity of Montreal
1757	Elizabeth Inchbald. Actress
1758	Johann Dannecker. Sculptor
1767	Gabriel Richard. Missionary
1776	Cpt. Reuben Weekes
1815	Moritz Brasig. Composer
1818	Irvin McDowell. Union general of Civil War
1829	Asaph Hall. Astronomer

1830	Helen Jackson. Novelist
1831	John Creighton. Philanthropist to Creighton University
1832	Bartholomew Demoret
1836	Emma Lunkenheimer. Pioneer
1836	James Tissot. Painter
1839	Fanny Cadwallander
1857	Andrew Jackson Shipman
1856	Oscar Wilde. Author
1858	Adm. William Sims. World War I Naval commander; president of Naval War College
1861	John Belford. Editor
1865	Patrick Henry Callahan of National Child Labor Commission; first American to be honored as a Knight of the Order of St. Gregory the Great
1872	Edith Galt Wilson
1875	Lynn Dana. Pianist; composer
1878	Elizabeth Daly. Poet
1883	Adm. Robert Ghormley. Commander at Guadalcanal
1886	Jonas Ingraham. Commissioner of All American Football Conference
1892	Ina Claire. Actress
1908	Bruna Castagna. Contralto
1919	Chuck Stevenson. Race car driver
1920	Mario Puzo. Author
1924	Lee Iacocca
1926	Evan Hunter. Writer
1926	Jean Peters. Actress
1937	Linda Lavin. Actress

October 16

1351	Gian Visconti. Ruler of Pavia
1430	James II. King of Scotts
1483	Gasparo Contarini. Statesman
1548	Isabel Forbes of Aberdeen
1588	Luke Wadding. Historian
1742	Joseph Scott. Patriot
1743	Joshua Gist of Revolutionary War
1750	Gilbert Longstreet. Patriot
1758	Noah Webster. Editor of *The American Dictionary of the English Language*

1790	Enoch Tucker
1818	Ignatius Mrak. Missionary
1827	Arnold Bocklin. Painter
1836	Terence Quinn. Senator
1837	John Barnell. Composer
1838	Horace Scudder
1861	John Bagnell Bury
1888	Eugene O'Neill. Playwright
1919	Kathleen Windsor. Novelist
1921	Nolan Frizzelle. Legislator
1925	Angela Lansbury
1927	Gunter Grass. Writer

October 17

1697	Augustus III. King of Poland
1729	John Lukens of the Colonies
1729	Pierre Alexandre Monsigny. Composer
1752	Amaziah Cushman. Patriot
1781	Richard Mentor Johnson. Senator; Vice President of the United States
1812	Henrietta Plummer. Pioneer
1813	Georg Büchner. Playwright
1817	Alexander Mitchell. Banking pioneer; grandfather of General Billy Mitchell
1825	Sylvanus Cobb Hausen
1833	Hon. Martin Dewey
1833	Paul Bert
1843	Francis Janssens. Archbishop of New Orleans
1843	Sherwood Davidge
1857	Ethelrod Taunton. Writer
1859	George Wassom. Settler
1860	Jane Barlow. Writer
1876	John J. Anthony
1895	Doris Humphrey. Dancer
1908	Jean Arthur. Actress
1912	Pope John Paul I (Albino Luciano)
1918	Rita Hayworth. Actress
1923	June Allyson. Actress
1928	Junior Gilliam of baseball
1928	Julie Adams. Actress

1930   Beverly Garland. Actress
1933   William Anders. Astronaut
1938   "Evil" Knievel. Motorcycle stuntman

October 18

1405   Pope Pius II (Enea Silvio Piccolomini)
1415   Heinrich von Disson. Writer
1481   Richard Beauchamp. Bishop of Salisbury
1569   Marino Giambattista. Poet
1595   Edward Winslow. One of the founders of Plymouth Colony
1632   Luce Giordano. Artist
1706   Baldassare Galuppi. Composer
1707   Mehitabel Allen. Descendant of Priscilla and John Alden
1729   Darius Lobdell of the Colonies
1753   John Brodie. Patriot
1759   Zabadee Cutting of the Revolutionary War
1787   Robert Livingston Stevens. Shipbuilder
1817   Alexander Mitchell. Architect of the Chicago, Milwaukee & St. Paul Railroad
1824   John Noble Goodwin. Statesman
1824   Don Valera. Novelist
1839   Thomas B. Reed. Statesman
1849   William Matson of the shipping line
1853   Henry Semple. Author
1870   Daisetz Suzuki. Writer
1875   Adm. Harry Yarnell. Fleet commander and aviation leader; commander-in-chief of the Asiatic Fleet
1886   William Patrick O'Connor. First bishop of Madison
1889   Fannie Hurst. Novelist
1894   H.L. Davis. Writer
1900   Lotte Lenya. Singer
1901   Ida Kristeller of Bourdeaux
1926   Chuck Berry. Singer
1926   Klaus Kinski. Actor
1927   George Scott. Actor
1933   Ludovico Scarfiotti. Race car driver
1958   Thomas Hearns. Holder of five world titles

October 19

1433   Marsilio Ficino. Philosopher
1609   Giovanni Bona. Author

1669	Bl. Angelo of Acri
1686	Peter Basch. Writer
1720	John Woolman. Patriot
1733	John Lowell of the Colonies
1774	Jean Faribault. Trader
1784	Leigh Hunt. Writer
1784	John McLoughlin. Pioneer
1810	Cassius Marcellus Clay. Founder of Kentucky's Emancipation Party
1826	John Quinlan. Bishop of Mobile
1833	Adam Gordon. Jockey
1834	Gen. Francis Channing Barlow of Union Army
1861	William Burns. Detective
1862	Auguste Lumiere. Inventor
1876	Mordecai Brown of St. Louis Cards
1832	Umberto Baccioni. Painter
1886	Thomas Mason. Painter
1885	Charles Driscole. Writer; authority on pirates and lost treasure
1895	Lewis Mumford. Author
1900	Erna Berger. Soprano
1901	Adm. Arleigh Burke. Chief of Naval Operations
1907	Ralph Bass. Writer
1917	Walter Munk. Geophysicist; oceanographer
1925	Zoot Sims. Saxophonist
1937	Peter Max. Artist
1967	Amy Carter

October 20

740	St. Acca. Bishop of Hexham
1496	Claude. First Duke of Guise
1547	Ven. Ursula Benincasa
1554	Bálint Balassa. Poet
1632	Sir Christopher Wren. Astronomer; architect
1644	Philip Eastman
1674	James Logan. Mayor of Philadelphia; Chief Justice of Pennsylvania Supreme Court; scholar; writer
1711	Gen. Timothy Ruggles of French and Indian Wars
1712	Gregor Zallwein
1759	Chauncey Goodrich. Legislator; congressman
1772	Samuel Coleridge. Poet

1785	Daniel Drake. Established medical school and other educational institutions
1816	James Wilson Grimes. Statesman
1825	Daniel Sickles. Military leader of Civil War; congressman; military governor of the Carolinas
1837	David Neal. Painter
1859	John Dewey
1890	Sherman Minton. Senator; Supreme Court Justice
1874	Charles Ives. Composer
1878	Upton Sinclair
1882	Bela Lugosi. Actor
1905	Manfred Lee. Co-author of Ellery Queen Mystery stories
1913	Grandpa Jones. Country singer; banjoist; fiddler
1925	Art Buchwald. Columnist
1931	Mickey Mantle
1934	Empress Michiko of Japan
1937	Wanda Jackson. Singer
1938	Juan Merichal of Baseball Hall of Fame
1953	Keith Hernandez of baseball
1955	Aaron Pryor. Welterweight champion

October 21

1349	William Bardolf. Fourth Lord of Wormgay
1581	Domenico Zampieri
1621	Nicolas Barre
1650	Jean Bart. Settler
1681	William Gooch. Governor of Virginia
1730	Thomas Branch Wilson of the Colonies
1768	Henry Beste. Author
1772	Alexandre Choron
1775	Giuseppe Baini. Composer
1812	Gordon Newell Mott. Noted Jurist
1818	John Dalgairns. Writer
1818	Enoch Fitch Burr. Author
1833	Alfred Nobel. Founder of Nobel Prize
1837	Gen. James Beaver of the Union Army
1842	Daniel Starns. Poet
1848	Frank Gregory. Artist
1857	Frederick Oldenbach. Seismologist; meteorologist
1860	Thomas Murray. Inventor

1861 Frederic Remington. Artist
1872 Walter Burns. Author
1893 Orland Armstrong. Congressman
1896 Ethel Waters. Singer
1917 Dizzy Gillespie. Jazz composer
1928 Whitey Ford
1929 Ursula Le Guin. Writer
1937 Michael Landon
1944 Jean Adelsman. Editor
1950 Ronald McNair. Astronaut

October 22

1071 William VII. Duke of Aquitaine
1688 Nadir Shah. Defeated Ottoman Turks and captured Baghdad
1689 John V. King of Portugal
1759 Thomas Cooper. Patriot
1783 Constantine Rafinesque. Writer
1792 Horace Kelsey of Killingsworth
1811 Franz Liszt. Composer
1819 George Eliot. Novelist
1822 Collis Huntington. Railroad pioneer
1830 Michael Pantenberg. Pioneer settler
1833 James Gory. Postmaster General
1841 Annie Cary. Contralto
1843 Stephen Babcock. Modernized the dairy industry
1844 Sarah Bernhardt
1854 James A. Bland
1875 Harriet Adams. Explorer
1875 Christopher Becker. Missionary
1876 Cissie Loftus. Actress
1880 Joseph Carr. A founder of NFL
1881 Clinton Davisson. Scientist
1889 John Balderston. Journalist
1894 Mei Lan-Fang. Actor
1903 George Beadle. Nobel Prize winner
1905 Constance Bennett. Actress
1907 Jimmie Foxx of Baseball Hall of Fame
1917 Joan Fontaine. Actress
1944 John Wetzel. Professional basketball coach
1963 Brian Boitano. Skating champion

October 23

912	Otto I. Holy Roman Emperor
1745	Aaron Gales of the Colonies
1752	Nicolas Appert. Founded first commercial cannery
1769	James Ward. Painter
1790	Chauncey Allen. Lexigrapher; established the "Christian Quarterly Spectator"; author
1801	Benjamin Raymond. Pioneer
1805	John Bartlett. Historian
1805	Adalbert Stifter. Poet
1824	Charles Fechter. Actor
1831	Basil Gildersleeve. Scholar
1835	Adlai Stevenson. Vice President of the United States
1842	Martin Griffin. Journalist
1844	Wilhelm Leibl. Painter
1847	Henry Preserved Smith. Historian
1848	Joseph Tasse. Journalist
1850	Sophus Bauditz. Writer
1860	Molly Seawell. Novelist
1871	Gjergj Fishta. Poet
1880	Dominikus Bohm. Architect
1895	Clinton Anderson. Senator; Secretary of Agriculture
1899	Emily Kimbrough. Author of *Our Hearts Were Young and Gay*
1904	Victoria Lincoln. Writer
1906	Gertrude Ederle. Champion swimmer
1918	Paul Rudolph. Architecture
1920	Bob Montana. Creator of "Archie"
1925	Johnny Carson
1927	Barron Hilton
1958	Gina Capra. Musician

October 24

51	Domitian. Roman Emperor
1682	Pierre Charlevoix. Explorer
1737	Deliverance Armington of Rehaboth
1753	Marian Dobmayer. Pioneer
1760	Ebenezer Keyes. Patriot
1762	Gen. Henry Sewall of the Revolution
1788	Sarah Hale. Novelist

1792	Richard Keith Call. Railroad builder
1830	Belva Ann Lockwood. Presidential candidate
1837	Samuel Ely
1842	Eliza Cosey. Pioneer
1843	Nellie Barrett
1855	James Schoolcraft Sherman. Vice President of the United States
1861	Henry Poole. Writer
1869	Herman Bauerle. Composer
1869	Maggie Bruton. Writer
1877	Grace Kessler. Author
1882	Sybil Thorndike. Actress
1887	Queen Victoria Eugenia of Spain
1895	Pierre. Duke of Valentinois
1914	Clay Smith. Race car driver
1925	Luciano Berio. Composer
1939	Murray Abraham. Actor
1947	Kevin Kline. Actor

October 25

1338	Elizabeth De Segrave. Daughter-in-law of Crusader John de Mowbray
1692	Elizabeth Farnese. Queen of Spain
1743	Timothy Whitney of the Colonies
1760	Sylvanus Crowell. Patriot
1765	Barzilla Spear of Braintree
1785	Francis Beckman. Archbishop of Dubuque; founded Dubuque Symphony Orchestra, Columbia Museum and the National Antiquarian Society
1795	John Pendleton Kennedy. Editor
1800	Thomas Macaulay. Poet
1801	Richard Bonington. Painter
1825	Johann Strauss II. The "Waltz King"
1831	Heinrich Bruck. Historian
1838	Georges Bizet. Composer
1877	Henry Norris Russell. Astronomer
1888	Richard Byrd. Explorer of Antarctica
1890	Floyd Bennett. Aviator of Greenland Expedition
1891	Fr. Charles E. Coughlin. Famous radio priest to forty million listeners; founder of the Shrine of the Little Flower; expert on money, labor and family; champion of social justice

1912	Minnie Pearl
1914	John Berryman. Poet
1918	Albert Brown. Mayor of Riverside
1928	Anthony Franciosa. Actor
1932	Glenn Gould. Pianist
1934	Jerry Patterson. Mayor of Santa Ana
1935	Russell Schweickart. Astronaut
1941	Anne Tyler. Novelist
1959	Lawrence Allen. Guitarist

October 26

1466	Desiderius Erasmus. Classical scholar
1556	Ahmad Baba. Writer
1625	Diego Altamirano
1716	Col. Benadam Gallop
1685	Demonico Scarlotti. Composer
1753	Charles Lining. Patriot
1757	Asher Robbins. Senator
1796	James Curley. Astronomer
1798	Beda Weber. Composer
1818	Elizabeth Prentiss. Writer
1830	Oska Joger. Historian
1833	Adelaide Phillipps. Actress
1855	George Burton. Inventor
1868	Samuel Prentiss Baldwin of Hillcrest Farm
1868	George Battle
1882	John Madden. Governor of New York Stock Exchange
1888	Patrick Byrne. Missionary
1900	Karin Boyle. Poet
1911	Mahalia Jackson
1914	Jackie Coogan. Actor
1919	Edward Brooke. Senator
1925	Stanley Adams. Publisher
1928	Albert Brewer. Governor of Alabama
1929	Georgette Amovitz. Choreographer
1938	Marcia Altman. Dancer
1942	Jonathan Williams. Race car driver

October 27

939	Athelston. King of Wessex
972	Robert the Pious of Orleans

1300	Baron Ralph Basset
1401	Catherine of Valois
1728	James Cook. The "Captain Cook" of the sea
1733	Benoni Lockwood of the Colonies
1736	James McPherson. Writer
1784	Giambattista Pianciano. Scientist
1800	Benjamin Wade. Senator
1811	Isaac Singer. Developer of sewing machine
1819	Henry Plant. Builder of railroads and steamship line
1827	Albert Fink. Railroad great
1833	Marie Paul Taglioni. Performer
1858	Theodore Roosevelt. President of the United States
1869	Viola Allen. Actress
1875	Janet Scudder. Painter
1910	Jack Carson. Actor
1914	Dylan Thomas. Poet
1917	Marshall Goldberg of National Foundation Football Hall of Fame
1924	Ruby Dee. Actress
1928	Kyle Rote. Sportscaster
1937	Ken Schultz. Mayor of Albuquerque
1939	Dallas Frazier
1940	Pamela McCorduck. Writer
1946	Peter Martins. Ballet master
1953	Peter Firth. Actor
1955	Susan Padilla Weingart

October 28

1017	Henry III. King of Germany
1430	Sir Robert West
1510	St. Francis Borgia
1550	St. Stanislas Kostka
1599	Ven. Marie de Incarnation. Missionary to Canada
1623	Johann Grueber. Missionary
1696	Maurice Saxe. Author
1789	Levi Coffin. "President of the Underground Railroad"
1796	John Law. Judge; congressman
1803	Caroline Unger. Contralto
1826	Homer Martin. Painter
1840	Joseph Fifter

1844	Moses Ezekiel. Sculptor; did busts of Washington, Lee and Jefferson
1846	Georges Auguste Escoffier. Chef
1856	John Coehne. Mayor of Evansville
1868	James Connolly. Author
1870	George Seropian of the five Seropian brothers who were the first Armenian settlers of Fresno
1894	Robert Murphy. Undersecretary of State; author
1902	Elsa Lancaster
1903	Evelyn Waugh. Writer
1904	George Dangerfield
1919	Walter Hansgen. Race car driver
1926	Bowie Kuhn. Baseball Commissioner
1927	Cleo Laine. Vocalist
1933	Edward H. Sothern. Actor
1949	Bruce Jenner. Track and field star; sportscaster

October 29

1508	The Duke of Alva
1711	Laura Maria Bassi
1729	Hezekiah Lovejoy of the Colonies
1740	James Boswell
1745	Thomas Lee. Governor of Maryland
1775	Jean Faribault. Trader
1809	Joshua Moody Young. Bishop of Erie
1817	John Gavit. Engraver; president of American Bank Note Company
1817	Jacob Davis. Pioneer
1846	Lamuel Amerman of Congress
1850	Emiluis Oviatt Randall. Historian
1855	Louis Bruchesi
1875	Alva Adams. Senator
1875	John Sigstein. Missionary
1884	Winifred Louise Ward. Writer
1891	Fanny Brice. Created "Baby Snooks" of radio
1893	Anna Case. Singer
1916	Joselyn Gill. Astronomer
1921	Bill Mouldin. Cartoonist
1926	Jon Vickers. Tenor
1927	Frank Sedgman. Tennis champion

1941   Jody Miller. Singer
1948   Kate Jackson. Actress
1961   Randy Jackson. Singer

October 30

1735   John Adams. Signer of the Declaration of Independence; first Vice President and second President of the United States
1741   Angelica Kauffmann. Artist
1751   Valentine Libbart
1762   André de Chénier. Poet
1773   Hugh White. Statesman; senator
1786   Philippe Gaspé. Novelist
1789   Hiram Bingham. Missionary
1799   Ignace Bourget
1825   Adelaide Proctor
1854   John Young. Musician
1860   Frederic Bancroft. Author
1864   Elizabeth Sprague Coolidge. Pianist: soloist
1867   Louis Austin. Scientist
1871   Paul Valéry. Poet
1881   Elizabeth Robert. Poet
1882   William Halsey. Admiral of the Fleet
1888   Alan Kirk. American Naval Commander at the D-Day Normandy landing
1890   Edward Doherty. Journalist
1890   Francis Monaghan
1893   Charles Atlas
1895   Dickinson Richards. Scientist
1896   Ruth Gordon. Actress
1908   Nino Farina. Race car driver
1914   Patsy Montana. Singer
1914   Marion Ladewig. Championship bowler
1917   Maurice Trintignant. Winner of 1938 and 1939 Grand Prix at Chimay; winner La Faucille hillclimber

October 31

1291   Philippe Vitry. Composer
1357   Sir Robert de Ferrers of England

1538	Ven. Caesar Baronius. Historian
1621	Richard Bartlett. Early settler
1684	Nathaniel Smith
1705	Pope Clement XIV (Giovanni Ganganelli)
1740	Philippe Lou Therbourg. Painter
1751	Richard Sheridan. Playwright
1758	Zachariah Allbough
1790	William Branford Shubrick. Mexican War leader and commander of Paraguay expedition of 1859
1795	John Keats. Poet
1796	Walter Scott
1805	Francis Weninger. Missionary
1826	Hugh Ewing. Fought in Civil War; U.S. Minister to the Hague; writer
1830	Adolf Baeyer. Nobel Prize recipient
1831	Alexander Randall. Postmaster General
1835	Adelbert Ames. Recipient of the Congressional Medal of Honor and Reconstructionist governor of Mississippi
1852	Mary Eleanor Freeman. Novelist
1860	Andrew John Volstead. Author of the Volstead Act (Prohibition)
1860	Reinhard Aschenbrenner
1869	William Moffett. Father of Naval Aviation
1872	Georges Barrere. Flutist
1874	Albert Trever. Historian
1883	Sara Allgood. Actress
1886	Chiang Kai Shek
1886	Commodore Thomas Council
1895	Forrest Percival Sherman. Chief of Staff of the Pacific Fleet
1897	Irving Hexter. Publisher
1912	Dale Evans
1924	Enrico Baj. Artist
1930	Michael Collins. Astronaut
1931	Lee Grant. Actress
1947	Frank Shorter. World class runner

November 1

1500	Benvenuto Cellini. Sculptor
1539	Pierre Pithou. Writer
1565	Mary Walter
1755	Jeremiah Dauchy. Patriot

1757	George Rapp of the American Revolution
1764	Stephen Van Rensselaer
1778	Gustave IV. King of Sweden
1801	Vincenzo Bellini. Composer
1814	Frances Whitcher. Humorist
1834	Marie Tancisius. Initiator of International Eucharistic Congress
1850	Louis Gratacap. Scientist
1864	Margaret Sherwood. Author
1871	Stephen Crane. Author of *The Red Badge of Courage*
1878	Eleanor Cisneras. Mezzo-soprano
1880	Grantland Rice. Sportswriter
1886	Sakutaro Haguwara. Poet
1894	Nelia White. Actress
1903	Max Adrian. Actor
1911	Celestine Damiano. Bishop of Cowden
1923	Gordon Dickson. Writer of science fiction
1923	Victoria De Los Angeles. Soprano
1929	Nicholas Mavroules. Mayor of Peabody; congressman
1935	Gary Player of golf
1937	Bill Anderson. Country Western singer
1960	Fernando Velenzuela of Los Angeles Dodgers

November 2

1299	Elizabeth Comyn of England
1734	Daniel Boone
1739	Karl Dittersdorf. Composer
1740	Samuel Hall. Founded first printing house in Salem
1749	Valentine Lewis of the Colonies
1755	Marie Antoinette
1788	Plyna Boyden
1795	James Polk. President of the United States
1799	Robert Turnbull
1808	John Ingham
1815	George Boole. Mathematician
1820	Lucretia Hale. Author
1821	George Rackley
1831	Henry Jones. Author
1848	Leslie Shaw. Governor of Iowa
1861	Maurice Blondel. Philosopher
1865	Warren Harding. President of the United States

1877	Aga Khan III
1880	Brian Hooker. Songwriter
1884	Thomas Welch. First bishop of Duluth; established the *Duluth Register*
1885	Harlow Shapley. Astronomer
1887	Thomas Kennedy. Labor leader
1892	Alice Brady. Actress
1903	Jack Oakie. Actor
1913	Burt Lancaster
1914	Johnny Vandermeer of Cincinnati Reds, Chicago Cubs and Cleveland Indians
1938	Pat Buchanan
1938	Queen Sophia of Spain
1954	John R. Lewis. Legislator

November 3

1610	Capt. George Curwen of Northhamptonshire
1723	Samuel Davies of the Colonies
1751	Joseph of Sandwich. Signer of the famous "Association Test" of 1776 in defense of the Colonies
1777	Deacon Seth Alden. Descendant of Priscilla and John Alden
1785	Charles Forbin-Janson. Founder of the Association of the Holy Childhood
1793	Stephen Austin. Commander of Texas Volunteers; Secretary of State of the Republic of Texas
1794	William Cullen Bryant. Poet
1799	William Sprague. Governor of Rhode Island
1801	Karl Baedeker. Publisher of Travel Guide Books
1816	Jubal Early. Confederate general
1818	James Renwick. Architect
1845	Vincenzo Bellini. Composer
1845	Edward Douglas White. Chief Justice
1850	James Mason. Statesman
1854	Jokichi Takamine. Scientist
1867	John Oliver Hobles
1867	Pearl Craigie. Novelist
1889	Heinrich Campondank. Painter
1890	Theodore Maynard. Author
1901	Leopold III. King of Belgium
1908	Branko Nagurski. All American

1909	James Reston. Journalist
1918	Bob Feller. Pitcher
1922	Charles Bronson
1933	Michael Dukakis. Governor of Massachusetts
1949	Larry Holmes. Heavyweight champion
1961	Viscount Linley. Son of Princess Margaret Rose of England

November 4

1575	Guido Reni. Baroque painter
1605	William Habington. Poet
1638	Lt. Nathaniel Holcombe
1714	Light Townsend
1758	Scriptur Frost
1759	Eleazer Crawford. Patriot
1760	Sherebian Fletcher
1774	Carlos Bustamonte
1806	Robert Launitz. Sculptor
1810	Lucius Robinson. Governor of New York
1816	Stephen Field. Supreme Court Justice
1826	Emmanuel Domeneck
1841	Benjamin Goodrich. Founder of B.F. Goodrich Company
1864	Anita McGee. Founder of the Army Nurse Corps
1869	Lucienne Breval. Soprano
1873	Bobby Wallace of St. Louis Cardinals and St. Louis Browns; managed St. Louis Browns and Cincinnatti Reds
1879	Will Rogers
1900	Don Page. Actor
1904	Walter Bauer. Prolific writer of prose, poetry, verse and juvenile works
1906	Bob Consideine
1906	Sterling North. Poet
1916	Walter Cronkite
1918	Art Carney
1918	Cameron Mitchell

November 5

1305	Robert de Clifford of Skipton
1715	Bl. Felix of Nicosia
1728	Franz Wulfen. Botanist

1744	Daniel Cushing. Patriot
1777	Filippo Taglioni of ballet
1779	Washington Allston. Painter
1796	Nathan Sanborn Page. Early settler of "Page Hill"
1821	Arthur Gilman. Architect
1832	John C. Malone. Justice of the Peace; Township Clerk; Sheriff of Scioto County, Ohio
1833	John Foley. Bishop of Detroit
1837	Arnold Janssen. Founder of the Society of the Divine Word
1850	Ella Wheeler Wilcox. Poet
1857	Ida Tarbell. Writer
1879	Will Hays of the Hays Office
1884	Joseph O'Mahoney. Senator
1885	Will Durant. Historian
1891	Greasy Neale. Coach at West Virginia Wesleyan, Washington and Jefferson, Marietta, Virginia and Yale
1892	Frederick Mayer. Editor
1895	Walter Griseking. Pianist
1904	Raja Rao. Novelist
1912	Roy Rogers
1921	Princess Fawzia
1941	Art Garfunkel
1963	Tatum O'Neill

November 6

1479	Joan, Queen of Castile. Mother of Emperor Charles V
1657	Joseph Denis of Three Rivers
1753	George Beaumont. Artist
1756	Richard Dale of the Navy
1797	Frances Drake. Actress
1831	Matilda Wood. Actress
1854	John Philip Sousa
1855	Paul Kalish. Opera singer
1860	Franklin Wheeler Mondell. Land office commissioner
1861	James Naismith. Inventor of basketball
1868	John Forsell. Baritone
1880	Robert Musil. Novelist
1883	Arnold Box. Composer
1887	Walter Johnson of Washington Senators

1890	Nicola Montani. Composer
1892	Sir John Alcock. Early aviator
1892	Harold Ross. Pioneer
1903	Elton Grenfell. Famous submariner
1920	Nicola Rossi-Lemini. Singer
1931	Mike Nichols. Director
1932	Stonewall Jackson. Singer; guitarist
1936	David Wart Stenman. Composer
1949	Brad Davis. Actor
1955	Maria Shriver

November 7

1515	John Newton of England
1604	Bl. Bernard of Offida
1731	Robert Rogers. Commander of Colonial Rangers in French and Indian War of 1754-63; commander at Makinac; writer
1741	Henry Hoas of the Revolutionary War
1741	Susanna Tisdale. Early settler of Lebanon, Connecticut
1758	Isaac Berruyer. Patriot
1763	Benedict Flaget. Missionary; bishop of Bardstown
1763	William Littlejohn
1795	Charles Denison Kingsbury of Watertown
1797	Commodore Silas Stringham. Union Commander; president Naval Retiring Board
1806	Elijah Smith. Pioneer
1828	Paul Baudry. Painter
1846	Ignaz Brüll. Composer
1847	Lotta Crabtree. Actress
1850	Alfred Nathorst. Geologist
1855	William Dorsey Jelks. Governor of Alabama
1867	Marie Curie. Physicist
1878	Dr. Lise Meitner of nuclear fission
1910	Pearl Argyle. Actress
1913	Albert Camus. Won Nobel Prize in 1957
1918	Billy Graham
1926	Joan Sutherland. Soprano
1928	Frances Fay
1942	Johnny Rivers. Singer
1943	Joni Mitchell. Singer

November 8

1656	Edmund Halley. Astronomer
1680	Constanzo Beschi. Missionary
1732	John Dickinson. Member of Continental Congress
1735	George Plater. Governor of Maryland
1758	Zaccheus Lovewell of the Colonies
1768	Hunking Penhallow. Early Virginian
1765	Martin Navarrete. Historian
1772	Charles Beauvois. Writer
1780	Samuel Foote. Governor of Connecticut
1789	Joseph Berard
1803	Henry Glassford Bell. Poet
1815	Mary Stephens Skinner
1818	Marco Minghetti. Statesman
1830	Oliver Howard. Union general
1849	Peter Funfsinn. Commissioner
1854	Johannes Rydberg. Mathematician
1883	Arnold Bax. Composer
1891	Neil Gunn
1898	Joseph O'Meara
1900	Margaret Mitchell. Author of *Gone With the Wind*
1904	Cedric Belfrage. Journalist
1909	Alberto Erede. Conductor
1909	Katharine Hepburn
1927	Patti Page
1934	Paul Wiggin of Cleveland Browns
1936	Edward Gibson. Astronaut

November 9

1664	Henry Wharton. Writer
1694	Marquard Hergott. Historian
1721	Mark Akenside. Poet
1731	Benjamin Banneker. Astronomer
1732	Mattias Schoenberg. Author
1734	Giuseppi Bertieri
1752	Philip Grandidier. Historian
1753	Christopher Lugar. Patriot
1795	Josiah Tattnall. Captain of Confederate Navy
1801	Gail Borden. Inventor of process for condensing milk

1803	Henry Farnjham. President of Chicago & Rock Island Railroad
1816	Joseph Bayma. Mathematician; textbook author
1819	Cornelius Blair
1820	Charles Hathaway Larrabee. Member of Wisconsin Constitutional Convention
1825	Gen. Ambrose Hill of the Confederate Army
1836	David Gray. Journalist
1841	Edward VII. King of England
1854	Maud Howe Elliott. Biographer
1869	Marie Dressler
1870	Auguste Tack. Painter
1887	Gertrude Astor. Actress
1896	Clifton Webb. Actor
1918	Florence Chadwig. Set swimming record
1918	Spiro Agnew. Vice President of the United States
1921	Jerome Hines. Singer
1925	Alistair Allan Horne. Author
1928	Anne Harvey. Poet
1934	Carl Sagan. Astronomer
1935	Bob Gibson. Pitcher

November 10

1341	Henry de Percy. Earl of Northumberland
1483	Martin Luther. Leader of the Protestant revolt
1559	Christop Brower. Historian
1592	Thomas Newberry of Yarcombe; came to Dorchester in 1634
1632	Samuel Allen of Braintree
1668	François Couperin. Composer
1683	George II
1728	Oliver Goldsmith. Writer
1758	John Lumsden. Patriot
1791	Robert Hayne. Governor of South Carolina
1792	Samuel Nelson. Supreme Court Justice
1808	Lewis Levin. Congressman
1820	Silas Bent
1820	Herman Cohen. Composer
1852	Henry Van Dyke
1871	Sir Winston Churchill
1874	Donald MacMillan. Arctic explorer
1879	Vachel Lindsay. Poet

1883	Arthur Ficke. Novelist
1886	"Pug" Ainsworth. Commanded U.S.S. "Mississippi" in World War II
1890	Claude Rains. Actor
1892	Frank Barrett. Congressman; state senator; U.S. senator; governor of Wyoming
1893	Mabel Normand. Comic
1893	John Phillips Marquard. Author
1903	Elizabeth Codell. Writer
1909	Birdie Tebbets of Detroit Tigers, Boston Red Sox and Cleveland Indians
1925	Richard Burton
1933	Ronald Evans. Astronaut
1949	Donna Fargo

November 11

995	Gisel of Swabia
1289	Nicholas Audley of Heleigh
1475	Pope Leo X
1642	André Boulle. Artist
1668	Johann Fabricius. Bibliographer
1693	Co. William Macon. Member of Virginia House of Burgesses
1729	Louis Bougainville. Explorer; discovered one of the Solomon Islands
1744	Abigail Adams. First Lady; notable for her letters which preserved much history of her time
1753	Adam Lu. Patriot
1799	Charles Bent. Pioneer
1821	Fëdor Dostoevski. Novelist
1833	Alexander Boridin. Composer
1836	Thomas Bailey Aldrich. Writer
1846	Anna Katharine Green. Author
1863	Paul Signac. Painter
1869	Victor Emmanuel III. King of Italy
1872	Maude Adams. Actress
1872	David Walsh. Senator; governor of Massachusetts
1882	King Gustav of Sweden
1883	Ernest Ansermet. Conductor
1885	Gen. George S. Patton, Jr. Commander of U.S. Third Army in World War II

1899	Pat O'Brien. Actor
1901	Richard Lindner. Painter
1907	George Templeton. Actor
1913	Robert Ryan. Actor
1914	Howard Fast. Novelist
1922	Kurt Vannegut, Jr. Science fiction writer
1925	Jonathan Winters. Actor
1928	Edward Zarinsky. Senator; mayor of Omaha
1937	Sam McQugg. Race car driver

November 12

1493	Baccio Bandinelli. Sculptor
1615	Richard Baxter
1648	Juana Ines de la Cruz. Poet
1655	Francis Nicholson of College of William and Mary
1684	Edward Vernon for whom Mount Vernon was named
1729	Francisco Alegre
1746	Jacques Charles. Constructed first hydrogen balloon
1809	Andrew McCormick
1814	Joseph Hooker. Commander in Civil War
1815	Elizabeth Cady Stanton
1836	Samuel David. Composer
1837	Michael Bass. Brewer
1840	François Auguste Rodin. Sculptor
1876	James Gillis. Missionary
1882	Giuseppe Borgese. Historian
1888	Anne Parish. Painter
1889	DeWitt Wallace. Co-founder of Reader's Direct
1891	Seth Nicholson. Astronomer
1892	Tudor Davis. Tenor
1893	Adm. Joseph Clark. Writer
1900	James Mitchell. Secretary of Labor
1906	George Dillon. Poet
1920	Richard Quine. Actor
1928	Bill Muncey. Winner of eight gold cups
1929	Princess Grace
1937	Richard Truly. Astronaut
1945	George Eaton. Race car driver
1949	Andre Laplante. Pianist

November 13

354	St. Augustine
1312	Edward III. King of England
1567	Maurice. Prince of Orange
1702	Dominic Vallarsi
1744	Zachariah Bunker. Patriot
1745	Valentin Haug. Founder of the first school for the blind
1750	Thomas Lindsay of Fairfax
1753	Ippolito Pindemonte. Poet
1755	Jonathan Tritle of the Colonies
1762	Pierre Bedard. Statesman
1782	Esaias Tegnér. Poet
1803	Adm. John Dahlgren of the capture of Savannah
1813	Zephyrim Engehardt. Missionary
1822	Eugene Casserly. Senator
1831	James Maxwell. Co-developer of Maxwell-Baltzmann kinetic theory
1838	Joseph Fielding Smith
1848	Honore Charles Grimaldi. Prince Albert I of Monoco
1850	Robert Louis Stevenson
1854	George Chadwick. Composer
1856	Justice Louis Brandeis
1886	Abraham Flexner. Biographer
1886	Mary Wigman. Dancer
1893	Edward Doisy. Nobel Prize winner
1903	Thomas Radall. Author
1906	Hermione Baddley. Actor
1909	Gunnar Bjornsbrand. Actor
1913	Alexander Scourby. Actor
1916	Jack Elam. Actor
1917	Robert Sterling. Actor
1933	Harold Lee Alexander. Author
1933	Adrienne Carri. Actress
1936	Robert Gonzales. Civic leader
1938	Jean Seberg. Actress
1942	John Paul Hammond. Guitarist
1946	Roy Hubbard. Singer
1947	Ket Wagner. Scientist
1948	Sheila Frazier. Actress

November 14

1454	Jeanne Hochette
1590	Gerrit van Honthout. Dutch painter
1601	St. John Eudes
1653	Jean Saint-Vallier. Bishop of Quebec
1765	Robert Fulton. Inventor of steam boat
1668	Johann Hildebrandt. Architect
1679	Omobono Stradivari. Violin producer
1756	Silas Blinn. Patriot
1764	Moses Broadwell of the Colonies
1774	Gasparo Spontini. Composer
1776	Henri Dutrochet. Discovered osmosis
1778	Johann Hummel. Pianist; conductor; composer
1779	Charles Lyell. Geologist
1807	Auguste Laurent. Scientist
1820	Anson Burlingame of Congress
1824	Joseph Fabre
1833	Henry Clay Barnabee. Actor
1838	Gen. Alvred Bayard Nettleton of Civil War
1840	Claude Monet. Painter
1861	Frederick Turner. Historian
1863	Leo Baekeland. Inventor
1893	Grant Illion Butterbough of Merrillan
1885	Constance Rourke. Writer
1900	Aaron Copeland. Composer
1909	Joseph McCarthy. Circuit judge; World War II marine officer; Senator from Wisconsin
1914	Eric Crozier. Writer
1921	Brian Keith. Actor
1924	Jean Madeira. Singer
1933	Fred Haise. Astronaut
1944	Mary Frances Casey. Actress
1934	Brett Lunger. Race car driver
1948	Prince Charles

November 15

1397	Pope Nicholas V (Thomas Parentricelli)
1486	Johann Eck
1648	Juan Salvatierra. Missionary

1692	Eusebius Amort
1708	William Pitt the elder
1738	William Herschel. Discovered planet Uranus
1748	Alexander Walden. Patriot
1750	Amosa Learned of Constitutional Convention
1784	Frances Allen. Daughter of Ethan Allen. First woman born in New England to enter religious life
1787	Eliza Leslie. Author
1807	Peter Burnett. First governor of California
1815	John Banvard. Painter
1827	John Merriott
1842	Charles Monroe Dickinson. Presidential elector
1849	James O'Neill. Actor
1856	Herbert Thurston. Writer
1874	Charles Merriam
1882	Felix Frankfurter. Supreme Court Justice
1884	Charles O'Donnell. Poet
1887	Marianne Moore. Writer
1891	Field Marshall Erwin Rommel. The "Desert Fox"
1891	William Vincent. Actor
1906	Chester Earl Morrow. Congressman; news commentator
1929	Ed Asner
1932	Petula Clark. Actress
1977	Peter Phillips. Son of Princess Anne of England

November 16

42BC	Tiberius Julius Caesar Augustus. Second Emperor of Rome
1459	Bl. Andrew of Rinn
1639	William Reed. Early settler
1684	Allen Bathurst
1693	Ebenezer Hopkins of the Colonies
1753	James McHenry. Secretary of War
1759	Job Loree. Patriot
1763	Josiah Barker. Shipbuilder
1764	Return Jonathan Meigs. Postmaster General
1766	Rodolphe Kreutzer. Violinist
1786	William Applegate of Congress
1831	Congressman Clinton Babbitt
1840	Jules Danbe. Violinist
1851	Joseph Burton. Senator

1851	Minnie Hauck. Soprano
1869	Frank Nankivell. Cartoonist
1872	Joseph Lynch. Bishop of Dallas
1873	William Handy. "Father of the Blues"
1889	George S. Kaufman. Playwright
1892	Harold Ross. Founder of *New Yorker Magazine*
1895	Paul Hindesmith. Composer
1895	Michael Arlen. Playwright
1896	Joseph Connor. Composer
1896	Laurence Tibbett. Baritone
1909	Burgess Meredith. Actor
1922	Royal Dano. Actor
1967	Lisa Bonet. Actress

November 17

1503	Angelo Bronzino. Artist
1685	Pierre La Vérendrye. Fur trader; explorer
1717	Jean Alembert. Developer of calculus
1734	Amos Yaw
1751	James Archer
1761	Charles Waterman
1771	Leopold Ackerman. Writer
1783	Anton Gunther
1786	William Barton. Botanist
1790	August Möbius. Astronomer
1812	William Warren. Actor
1858	Giuseppe Campanari. Baritone
1861	Wallace Nutting. Painter
1876	Homer Lea. Military genius; author
1878	Grace Abbott
1884	Robert Armstrong. Founder of *Sacramento Register*
1887	Viscount Montgomery. British field marshall
1897	Herbert O'Conor. Governor of Maryland
1902	Eugene Wigner. Developer of Nuclear Power
1904	Mischa Aver. Actor
1925	Mary Frances Francis. Editor
1930	Bob Mathias. Winner of Olympic Decathlon; congressman
1930	Mari Aldon. Actress
1935	Toni Sailer. Olympic skier
1938	Gordon Lightfoot

1944 Danny De Vito
1944 Tom Seaver of baseball

November 18

9 Titus Flavius Vespasianus. Roman Emperor
1680 Jean Baptiste Loeillet. Composer
1727 Isaac Judd of the Colonies
1756 Jonas Sams of the Colonies
1786 Karl Maria von Weber. Composer
1787 Louis Daguerre. Developer of modern photography
1788 Seth Boyden. Inventor
1810 Asa Gray. Botanist
1818 Samuel W. Austin of Crawfordsville
1818 James Renwick. Architect; designer of St. Patricks Cathedral in New York City
1847 William Norris. Novelist
1851 Robert Wynne. Postmaster General
1860 Ignace Paderewski. Pianist; composer; statesman
1871 Jesse Bonstelle. Actress
1871 Robert Hugh Benson. Author
1874 Carrie White
1874 Clarence Day. Humorist
1877 Valentine Breton. Writer
1882 Wyndham Lewis. Architect
1882 Amelita Galli-Curie. Soprano
1885 Fr. Joseph Kentenich. Founder of Schoenstatt
1888 Otto Erhardt. Producer; director
1891 Gio Ponti. Architect
1899 Eugene Ormandy. Conductor
1901 George Gallup. Opinion researcher
1908 George Wald. Winner of Nobel Prize
1908 Imogene Coca. Actress
1909 Johnny Mercer. Singer
1923 Rear Adm. Alan Shephard. Astronaut; first American in space
1942 Linda Evans

November 19

1190 Baldwin. Archbishop of Canterbury
1600 Charles I. King of England

1722	Leopold Auenbrugg. Inventor
1749	Levi Loveland
1752	Gen. George Rogers Clark of American Revolution
1758	Asa Blood. Patriot
1760	Sarah McFarland. Had 446 descendants when she died
1805	Ferdinand Marie Lesseps. Builder of Suez Canal
1805	Martin Kundig
1831	James Garfield. President of the United States
1831	Rose Whitty
1835	Gen. Fitzhugh Lee
1839	Atticus Haygood. Chaplain in Conferate Army; president of Emory College; editor
1844	Joseph Cotter. First bishop of Winona
1862	Billy Sunday. Pro baseball player; revivalist
1884	Augustin Fliche. Historian
1886	Hugh Allen. Actor
1888	José Capablanca. Chess genius
1894	Phyllis Bentley. Writer
1899	Allen Tate. Poet
1905	Tommy Dorsey. Bandleader
1915	Earl Sutherland. Winner of Nobel Prize
1917	Mrs. Indira Gandhi. Prime Minister of India
1919	Moise Tshombe. Statesman; provincial president of Katanga
1919	Roy Campanella. Baseball great
1926	Jean Kilpatrick
1937	Dick Cavett. Actor
1938	Lothar Matschenbacker. Race car driver
1950	James Adler. Pianist

November 20

1180	St. Edmund Rich
1579	Anne Mynne Calvert
1717	Col. Jacob Blackwell. Member of provincial convention
1726	Oliver Wolcott. Signer of the Declaration of Independence
1738	Girtan Walker
1744	Matthew Barnes of the Colonies
1761	Pope Pius VIII (Francesco Castiglione)
1772	Antipas Jackson
1807	Augustus Thebaud. Prolific writer; first president of Fordham University

1829	W.F. Rhoads. Saddler
1861	James Phelan. Senator
1867	Patrick Cardinal Hayes. Archbishop of New York
1869	Clark Griffith of baseball
1880	Fred Zimmerman. Governor of Wisconsin
1887	Ernest Hooton. Anthropologist
1888	Elsie Albert. Actress
1896	Robert Armstrong. Actor
1900	Chester Gould. Creator of Dick Tracy
1904	Alexandra Danclava. Ballerina
1908	Alistair Cooke. Journalist
1910	George Devine. Actor
1913	John Frederick Nums. Poet
1916	Judy Canova. Actress
1917	Bobby Locke of golf
1920	Gene Tierney. Actress
1943	Meredith Monk. Composer
1948	Gunnar Nillson. Race car driver; winner of Belgian Grand Prix
1956	Bo Derek. Actress
1963	Maldoris Phillips

November 21

1567	Ven. Anne de Zainctonge. Foundress of the Society of the Sisters of St. Ursula of the Blessed Virgin
1643	Robert LaSalle. Fur trader; first European to navigate the Mississippi River
1677	Flowe Norton. Pioneer
1725	Leopoldo Caldani. Anatomist
1729	Jonah Barrett. Signer of the Declaration of Independence
1733	Gen. Philip Schuyler of Continental Army. Senator
1733	Shattuck Blood. Patriot
1750	Owen Zebley
1750	Gideon Cushman. Patriot
1789	Cesare Balbo. Statesman; historian
1835	Hetty Green
1839	Gen. Andrew Sheridan Burt. Writer
1854	Pope Benedict XV (Giacomo della Chiera)
1873	Henry Arens. Congressman
1876	Olav Duun. Novelist
1891	Edward Ellsberg. Writer

1904 Coleman Hawkins. Jazz musician
1907 Jim Bishop. Newspaperman
1912 Eleanor Powell. Actress
1916 Sid Luckman of football
1916 Homer Abele. Congressman
1920 Stan Musial of baseball
1924 Vivian Blaine. Actress
1944 Mary Carsey. Producer

November 22

1515 Mary of Guise. Queen of Scotland; mother of Queen Mary Stuart
1710 Wilhelm Bach. Organist
1754 John Chenault. Patriot
1754 Abraham Baldwin
1767 Andreas Hofer
1800 Lynn Boyd. Legislator
1801 Charles Pise. Writer
1801 Joseph Plumpe. Editor
1806 Hugh Grigsby. Historian
1829 Shelby Moor Cullom. Pioneer
1830 Justin McCarthy. Historian
1857 George Gissing. Writer
1859 Cecil Sharp. Composer
1868 John Nance Garner. Vice President of the United States
1869 Frederic Washburn
1877 Endre Ady. Poet
1883 Ruby Davy. Pianist; composer
1889 Thomas Beer. Writer
1890 Charles De Gaulle
1898 Sarah Gibson Blanding. President of Vassar College
1899 Hoagy Carmichael. Composer
1904 Paul Bussard. Editor
1909 Wilma Pitchford. Author
1913 Benjamin Britten. Composer
1918 Henryk Szeryng. Violinist
1922 Lynne Roberts. Actress
1924 ·Geraldine Page
1925 Gunther Schuller. Composer; conductor
1930 Owen Garriott. Astronomer
1932 Robert Vaughn. Actor

1950 Lyman Bostock of California Angels and Minnesota Twins
1958 Jamie Lee Curtis

November 23

1221 Alfonso X. King of Castile
1505 Ercole Gonzaga
1637 Nehemiah Palmer of Stonington
1723 Jeremiah Cloud
1726 Edward Bass. Episcopal bishop; great-great-grandson of Priscilla and John Alden
1732 Thomas Cutts of the Colonies
1746 Thomas Bloomfield
1749 Edward Rutlege. Signer of the Declaration of Independence
1749 Benjamin Chittendon
1760 Isaac Anderson. Congressman
1797 Benjamin Hale
1804 Franklin Pierce. President of the United States
1821 Charles Maryon. Etcher
1840 Leopold Beaudenom. Writer
1855 Adm. Frank Friday Fletcher. Directed landings at Vera Cruz
1856 Edward O'Dea
1859 William Bonney. "Billy the Kid"
1869 Valdemar Poulson. Invented "Telegraphone"
1871 George Frost Archer. Inventor
1876 Manuel de Falla. Composer
1878 Adm. Ernest King. Commander-in-chief United States Fleet
1887 Boris Karloff. Actor
1907 Jack Schaeffer. Writer
1908 Luke Short. Author of popular westerns
1915 Ellen Drew. Actress
1916 Emmet Ashford. Actor
1931 Edward Panelli

November 24

1414 Albert III Achilles
1713 Fr. Junipero Serra. Builder of missions
1729 Adm. Pierre Estain. Governor of the Antilles
1748 Archibald Crary. Patriot

1784	Zachary Taylor. President of the United States
1800	Henry Oliver. Composer
1810	William Armstrong
1821	Henry Buckle. Historian
1828	George Sala. Journalist
1828	William Oraban. Author
1840	Charles Perkins. President of Chicago, Burlington & Quincy Railroad; president of Hannibal & St. Joseph Railroad; president of Kansas City, Coundil Bluffs & St. Joseph Railroad
1848	Lilli Lehmann. Soprano
1849	Frances Burnett. Author
1854	Walter George Smith. President of the American Bar Association; member U.S. Indian Commision; writer
1868	Scott Joplin. Composer
1869	Pliny Goddard
1876	Walter Griffin
1878	Sir Roger Backhouse. Admiral of the Fleet
1888	Dale Carnegie
1895	Rene Maison. Singer
1912	Teddy Wilson. Jazz pianist
1914	Geraldine Fitzgerald
1917	Howard Duff
1918	Helga Sandburg. Novelist
1925	Barbara Ledbetter. Author
1938	Oscar Robinson of Basketball Hall of Fame

November 25

1386	Bl. Elizabeth of Reute
1556	Jaques Dupperson
1562	Lope Vega. Playwright
1647	Samuel Gridley. Proprietor of Farmington
1664	Priscilla Hazen
1677	Andres Blanqui. Architect
1754	William Wilson. Patriot
1757	Henry Livingston. Supreme Court Justice
1774	Bl. Stefano Bellesini
1781	John Miller. Congressman; governor
1792	Robert Abell. Missionary
1816	Thomas Grant. First bishop Southwark

1835	Andrew Carnegie
1842	Rear Adm. Alexander Berry Bales
1846	Carrie Nation
1850	Charles Turner. Painter
1852	James Finney McElroy. Inventor
1859	Jane Hading. Actress
1862	Ethelbert Nevin. Composer
1870	Solanus Casey. Renowned for his ministry to the sick and needy
1881	Pope John XXIII (Angelo Roncalli)
1895	Wilhelm Walter Friedrich Kempff. Composer; pianist
1896	Virgil Thomson. Composer
1902	Eddie Shore of hockey
1904	Helen Japson. Singer
1912	Alvin Nikolois. Composer
1914	Joe Di Maggio
1920	Ricardo Montalban
1924	Patrick Minor Martin
1933	Kathryn Crosby
1941	Tina Turner
1963	Bernie Kosar of football

November 26

1617	Cornelius Hazart. Writer
1712	William Zane of the Colonies
1729	Isaac Luke. Patriot
1731	William Comper. Poet
1756	Christian Meza. Author
1767	Leopold Cicognara. Writer
1790	Peter Force. Historian
1792	Sarah Moore Grimke
1800	Casimir Ubaghs
1811	Alexander Hubner
1807	William Mount. Painter
1830	Horace Taber
1837	John Newlands. Scientist
1858	Mother Katharine Drexal. Founder of the Sisters of the Blessed Sacrament
1858	Israel Abrahams. Scholar; author; editor
1869	Queen Maud of Norway

1871	Frad Tenney. Fielder; catcher; manager of Boston Braves
1896	Helen White. Author
1907	David Ewen. Author
1912	Eric Sevareid. News commentator
1919	Frederick Pohl. Writer
1922	Charles Schulz. Cartoonist; creator of "Peanuts"
1923	Pat Phoeniz. Actress
1933	Robert Goulet. Actor
1935	Marian Mercier. Actress
1946	John McVie of Fleetwood Mac

November 27

1595	Alessandro Algardi. Sculptor
1701	Anders Celsius. Inventor of centigrade temperature scale
1703	James De Lancey. Lt. governor
1737	Jonathan Currier. Patriot
1756	Increase Sumner. Governor of Massachusetts
1763	Tunis Quick of the Colonies
1765	St. Joan Antida Thouret
1767	Gen. Bennett Riley of War of 1812
1806	Joseph Stevenson. Architect
1807	Abraham Van Buren of Mexican War
1809	Fanny Kemble. Author
1820	Augustine Hewit. Editor
1830	Jean Henry Vignaud. Statesman; author
1838	Paul Bedjan. Missionary
1843	Cornelius Vanderbilt II
1862	Hugh Henry. Author
1874	Charles Beard. Historian
1881	Thomas Ignatius Parkinson of New York City Charter Commission
1889	Senator Carl Hatch
1899	John Allen. Congressman
1900	Leon Barzin. Orchestra conductor
1902	Sprague De Camp. Author
1909	James Agee. Poet
1912	David Merrick
1942	Jimi Hendrix. Singer
1944	Eddie Rabbit

November 28

- 1632   Jean Baptiste Lully. Composer
- 1729   Jean Estaing
- 1748   Marmaduke Stone
- 1757   William Blake. Artist
- 1784   Christian Remsberg
- 1808   Samuel Barron. Youngest midshipman ever accepted in the U.S. Navy; commander of Confederate Naval forces
- 1811   Maximillian II. King of Bavaria
- 1821   Samuel Anable of the Army of the Potomac
- 1822   George Ellis Pugh. Senator
- 1829   Anton Rubinstein. Composer
- 1831   John McKay. Telecommunication pioneer
- 1831   Roberts Bartholow. Author
- 1835   S.G. Soverbill of the War of 1812
- 1857   King Alfonso XII of Spain
- 1868   Hermann Hahn. Sculptor
- 1893   Fairfax Downer. Author
- 1895   José Iturbi. Pianist; conductor; composer
- 1903   James McGrath. Governor of Rhode Island
- 1907   Alberto Moravia. Novelist
- 1909   Rose Bampton. Soprano
- 1912   Morris Louis. Painter
- 1915   Yves Theriault. Author
- 1932   Parry O'Brien. Track star
- 1935   Randolph Stow. Novelist
- 1959   Steven Roche. Cyclist

November 29

- 1338   Lionel of Antwerp
- 1627   John Ray. Scientist
- 1696   Anne Madeleine Remuzat
- 1728   Peter Powers
- 1729   Charles Thompson. Secretary of Continental Congress
- 1741   Jonah Hooker of the Colonies
- 1752   Jemima Wilkinson. Pioneer
- 1754   Daniel Toll Clute. Patriot
- 1760   Seth Blair
- 1760   Samuel Walker

1790	Daniel O'Shea
1797	Gaetano Donizetti. Composer
1824	Gen. Thomas Henderson. Collector of Internal Revenue; presidential elector
1829	Albert Bellows. Painter
1832	Louisa May Alcott. Author of *Little Women*
1837	Mariah Page
1849	Ambrose Fleming. Scientist
1879	Mary Spangler
1898	C.S. Lewis. Prolific author
1902	Mildred Harris. Actress
1908	Adam Clayton Powell of Congress
1917	Merle Travis. Singer
1923	Frank Reynolds
1927	Vince Scully. Sportscaster

November 30

538	St. Gregory of Tours
1318	Sir Henry FitzRoger of Somerset
1326	Baron Roger de la Warre
1466	Andrea Doria. Statesman; admiral of the Renaissance
1518	Andrea Palladio. Architect
1554	Philip Signey
1637	Louis Sébastien Tillemont. Historian
1667	Jonathan Swift. Author of *Gulliver's Travels*
1723	William Livingston. First governor of New Jersey
1729	Samuel Leabury of the Colonies
1744	Karl Knebel. Poet
1761	Smithson Tennant
1810	Oliver Winchester. Developer of Winchester rifles
1815	Isaac Newton Arnold. Congressman; biographer of Abraham Lincoln
1819	Cyrus Field
1823	Henry Miller
1835	Mark Twain
1868	Ernest Newman. Biographer
1875	Frank O'Malley
1878	Lucy Maud Montgomery. Author
1880	Richard Tawney. Economic historian
1890	Tracy Lewis. Writer

1895 Johann David. Composer
1913 Dorothy McGee. Historian
1919 Dr. Jane Cooke Wright. Pioneer in chemotherapy
1924 Shirley Chisholm of Congress
1929 Dick Clark. Entertainer
1931 Bill Walsh. Coach
1947 David Mamet. Playwright
1962 Bo Jackson of Kansas City Rebels

December 1

1083 Anna Comnena. Historian
1359 Agnes de Marten
1716 Étienne Falconet. Sculptor
1741 Henry Lumpkin. Patriot
1744 Epenetus Scofield
1756 Evan Truman of the Revolutionary War
1826 William Mahone. Railroader
1830 Matilda Heron. Actress
1844 Augustine Van De Vyver. Bishop of Richmond
1844 Alexandra Caroline, Princess of Schleswig-Hostein. Queen consort of King Edward VII of England; mother of King George V
1847 Julia Moore. Singer
1869 George Sterling. Poet
1874 Archbishop John Cantwell of Los Angeles
1875 Joseph Williams. Anthropologist
1879 Lane Bryant. Founder of clothing stores
1880 William Arndt. Writer
1886 Rex Stout. Detective story writer
1892 Mabel Leigh Hunt. Author
1901 Grace Moore. Singer
1910 Alicia Markova
1911 Walter Alston
1914 Mary Martin. Actress
1933 Violette Verdy. Ballerina
1939 Lee Trevino. Pro golfer
1940 Richard Pryor. Actor
1957 Patrick Bissell. Dancer

December 2

- 1563 Muzio Vitelleschi
- 1578 Agostine Agazzari. Composer
- 1694 William Shirley. Governor of Massachusetts
- 1738 Gen. Richard Montgomery of the Revolution
- 1743 Francesco Bianchi. Patriot
- 1749 Sunderland Sayre of the Colonies
- 1756 Stephen Fuller
- 1803 Robert Hawks. Pioneer
- 1813 Bishop Joseph Fessler. Secretary of the Vatican Council
- 1826 Bl. Maria Soledad Torres Acosta. Foundress of the Sisters Servants of Mary
- 1832 Michael O'Farrell. First bishop of Trenton
- 1832 Vincent de Paul Bailly. Journalist
- 1835 Edward Gruy. Author
- 1841 William Newton Clarke. Author
- 1857 James Allee. Senator
- 1859 Georges Seurat. Artist
- 1862 Florence Barclay. Novelist
- 1889 Paul Althouse. Tenor
- 1899 John Barbirolli. Conductor
- 1909 Helen Douglas Adam. Poet
- 1916 John Bentley. Actor
- 1917 Ezra Stan. Actor
- 1924 Gen. Alexander Haig. Secretary of State
- 1925 Julie Harris. Actress
- 1932 Bob Pettit of basketball

December 3

- 1368 Charles VI of France
- 1690 José Fanseca
- 1730 Daniel Averill of the Colonies
- 1731 Stefano Cardinal Borgia. Historian
- 1743 Benjamin Cutter. Patriot
- 1744 George Custer
- 1747 Joseph Vigo. Fur trader
- 1751 George Cabot of Constitutional Convention
- 1755 Gilbert Stuart. Novelist
- 1756 Sen. Aaron Ogden. Governor of New Jersey

1796	Francis Kenrick. Pioneer
1812	Hendrik Conscience. Novelist
1826	George McClellan. Union general
1829	Green Berry Raum. Congress
1834	Martin Morris. A founder of Georgetown Law School
1836	Frances Mosher
1838	Cleveland Abbe. Meteorologist
1848	Solomon Seelig
1862	Charles Fenstlemaker
1865	Gen. George Washington Burr
1871	Albert Johannsen. Retrographer
1871	Romanus Butin
1871	Newton Baker. Secretary of War
1900	David Cunningham. Bishop of Syracuse
1900	Richard Kuhn. Nobel Prize winner
1924	Roberto Mieres. Race car driver
1927	Phyllis Curlin. Soprano
1930	Andy Williams
1937	Bobby Allison. Race car driver
1964	Bobby Marshman. Race car driver

December 4

1584	John Cotton. Instrumental in effecting Roger William's banishment
1660	André Campra. Composer
1671	Henry Belsunce. Writer
1708	François Picquet. Missionary
1736	Thomas Godfrey. Playwright
1744	Jonathan Ford of the Colonies
1759	Caleb Carpenter. Patriot
1761	Christopher Raymond Perry. Naval hero
1768	Sarah Snell Bryant. Pioneer
1795	Thomas Carlyle. Historian
1809	Peter Garner. Abolitionist
1821	Wilhelm Temple. Astronomer
1835	Samuel Butler. Writer
1836	Marium Penn
1841	Henry Harrison Bingham. Awarded Congressional Medal of Honor for distinguished gallantry in the Battle of the Wilderness; served seventeen terms in Congress
1861	Lillian Russell

1864	Elizabeth Kite. Historian
1866	Vasili Kandinski. Originator of abstract art
1892	Francisco Franco. Decorated with the Cross of Maria Christina; youngest general in Europe; liberator of Spain
1893	Herbert Read. Poet
1910	Gregory "Pappy" Boyington. Flying Ace
1924	Robert Presley. Senator
1930	Harvey Kuinn of baseball
1937	Max Baer. Actor
1938	John Atkins. Pianist
1938	Rebecca Morgan. State senator
1963	Roger Ahl. Producer

December 5

1443	Pope Julius II (Giuliano della Rovere)
1661	Robert Harlow. First Earl of Oxford
1734	Jeremiah Atwater. Patriot
1756	Eliphalet Cheney of the Revolution
1757	Jean Ranzan
1782	Martin Van Buren. President of the United States
1822	Elizabeth Cabot Agozziz. First president of Radcliffe College
1830	Christina Rossetti. Poet
1833	Susan Hale. Artist
1841	Marcus Daly. Prospector
1856	Alice Brown. Author
1869	Ellis Butler. Author
1872	Frederick Murphy. Senator
1877	Charles Callan. Author
1887	Eliza Monroe "La Belle Americaine." Daughter of President Monroe
1889	Signey Algier. Actor
1890	Fritz Long. Director
1899	Vincent Sheean. Writer
1901	Walt Disney
1902	Strom Thurmond. Governor of South Carolina; Senator
1925	John Williams. Writer
1932	Jim Hurtuhise. Race car driver
1941	Jeffrey Miner Wallman. Author
1942	Richard Ackerman

December 6

345	St. Nicholas
1361	John de Sutton IV of Dudley Castle
1421	Henry VI. King of England
1478	Baldassare Castiglione
1554	Agustin Antolinez
1640	Claude Flewry. Historian
1748	Shikellamy. Oneida chief
1751	Jacob Longacre. Patriot
1761	Jean de Fates of the Colonies
1792	William II. King of the Netherlands
1802	Duncan Ingraham. Commander of Charleston Naval Station during Civil War
1806	Gilbert Duprez. Composer
1810	Victor Dechamps
1812	Hezekia Bateman. Actor
1833	John Mosby of the Confederacy
1857	Arnold Grantvoort. Missionary
1859	Edward Sothern. Actor
1870	Lucia Fuller. Painter
1870	William Hart. Actor
1884	James Kenney. Historian
1887	Lynn Fontaine. Actress
1886	Joyce Kilmer. Poet
1892	Lina Carstlus. Actress
1896	Ira Gershwin. Lyricist
1900	Agnes Moorehead. Actress
1901	Eliot Porter of photography
1904	Eve Curie. Writer; daughter of Pierre and Marie Curie
1920	George Porter. Scientist
1921	Otto Graham of football
1927	Adm. Donal Perry Hall. Recipient of Legion of Merit

December 7

521	St. Columbkill
1542	Mary Stuart. Queen of Scots
1545	Lord Darnley
1598	Giovanni Bernini Baroque. Artist
1604	Ambrose Corbi

1615	Nicodemus Tessin the Elder. Architect
1731	John Albertrandi
1747	Col. Ezekiel Polk of Revolutionary War
1761	Madame Toussaud of the wax museum
1801	Johann Nestroy. Singer
1802	Ludwig Lesser
1807	Hon. James J. Harrison
1832	Ella Edes. Correspondent
1859	Octaviano Larrozola. Governor of Rhode Island
1862	Henry Bemerunge. Historian
1864	Pietro Mascogni. Composer
1865	Sibyl Sanderson. Opera singer
1873	Willa Cather. Author of American West
1880	John Adam Wolfer of Macon County
1888	Matthew Brown. Author
1888	Matthew Broun, Jr. All American
1889	Gabriel Marel. Dramatist
1894	Stuart Davis. Painter
1897	Adm. Paul Heineman of World War II
1915	Eli Wallach. Actor
1923	Ted Knight. Actor
1928	Mickey Thompson. Race car driver; named "Fastest American on Wheels" in 1958
1930	Harold Mattson. Mathematician
1939	Heywood Brown. President of American Newspaper Guild
1949	Tom Waite. Singer

December 8

1626	Christina. Queen of Sweden
1708	Francis of Lorraine. Holy Roman Emperor
1758	William Breckenridge
1765	Eli Whitney. Inventor of the cotton gin
1792	Gioacchino Ventura di Paulica. Orator; patriot; philosopher
1815	Adolph Menzel. Illustrator
1819	James Mock
1821	Henry Poor. Editor
1828	Clinton Bowen Fisk. Founder of Fisk University
1828	Henry Timrod. Poet
1839	Alexander Cassatt. President of Pennsylvania Railroad
1840	Charles Herberman. Author

1861	William Crapo Durant. Founder of General Motors
1865	Jean Sibelius. Composer
1877	Frank Nichton. Railroad designer
1881	Albert Gleiszer. Artist
1886	Diego Rivera. Muralist
1889	Hervey Allen. Novelist
1913	Delmore Schwartz. Poet
1922	Jean Ritchie. Singer
1925	Sammy Davis, Jr.
1928	Jimmy Smith. Jazz organist
1933	Flip Wilson. Comedian
1937	James McArthur. Actor

December 9

1575	David Baker. Writer
1608	John Milton. Poet
1623	Ernest of Hesse
1720	Ebenezer Lovering. Patriot
1739	Buenaventura Siljar. Missionary
1742	Carl Scheele. Discovered oxygen
1850	Emma Abbott. Opera singer
1868	Fritz Haber. Scientist
1871	George Washington Ogden. Writer
1874	Joseph McSorley. Author
1882	Joaquín Turina. Composer
1886	Clarence Birdseye
1890	Charles Tansit. Historian
1894	Eddie Dowling. Actor
1898	Emmett Kelly. Clown
1899	Leoni Fuller Adams. Poet
1902	Margaret Hamilton. Actress
1902	Lucius Beebe. Journalist
1904	Louis Kronenberger. Drama critic
1905	Jamet Adam Smith. Author
1909	Douglas Fairbanks, Jr. Actor
1910	Gladys Zehnphennig. Novelist
1911	Broderick Crawford. Actor
1911	Lee J. Cobb. Actor
1915	Elizabeth Schwartzkoph. Opera singer
1917	James Rainwater. Nobel Prize winner

1918	Kirk Douglas. Actor
1919	William Lipscomb. Scientist
1922	Redd Foxx
1929	John Cassanetes. Actor
1942	Dick Butkus of football

December 10

1538	Giovanni Battista Guarini. Poet
1631	Francesco Lana. Mathematician
1656	Charles Le Moyne of battle against Iriquois
1745	Thomas Holcroft. Writer
1751	Amos Goff. Patriot
1770	James Hog. Writer
1795	Matthias Baldwin. Locomotive builder
1799	Joseph Cretin. Missionary; first bishop of St. Paul
1809	George Goldthwaite. Senator
1815	Henry Behrens. Missionary
1821	William Barry of Confederate Congress
1822	Cesar Franck. Composer
1824	George MacDonald. Writer
1826	John Kinkead. Governor of Nevada
1830	Emily Dickinson. Poet
1851	Melvil Dewey. Inventor of Dewey Decimal System
1869	Francis Donnelly. Writer
1870	Adolph Loos. Architect
1878	Cornelius McGlennon. Congressman
1888	Fritz Reiner. Conductor
1890	Edward Joseph O'Brien. Author
1910	Sy Oliver. Bullfighter
1913	Morton Gould. Composer
1940	Bentley Warren. Race car driver

December 11

1725	George Mason. Virginia statesman; George Mason University named for him
1731	Ahial Lovejoy of the Colonies
1736	Andrew Adams of Continental Congress
1750	Isaac Shelby. First governor of Kentucky

1757	Samuel Sewall. Legislator; judge
1781	Sire David Brewster. Optical pioneer
1803	Hector Berliz. Composer
1813	John Brouvillet
1855	Fernand Carral
1863	Annie Cannon. Astronomer
1866	George Summerton Parker. Inventor
1869	Annie Leslie. Journalist
1882	Fiorella La Guardia. Mayor of New York
1892	Ursula Bloom. Writer
1905	Gilbert Roland. Actor
1908	Elliot Carter. Composer
1910	Russell Ash. Actor
1913	Carlo Ponti
1924	Doc (Felix) Blanchard. "Mr. In" of Army
1931	Rita Moreno. Actress
1922	Grace Paley. Writer
1926	Willie Mae Thornton. Singer
1937	Jan Bach. Composer
1953	Bess Armstrong. Actress

December 12

1735	Peter Lozier. Patriot
1745	John Jay. First Supreme Court Chief Justice
1777	Emperor Alexander I
1779	St. Madeleine Barat. Foundress of the Sacred Heart Society
1786	William Marcy. Governor of New York
1790	Letitia Tyler. First Lady
1805	William Garrison. Abolitionist
1806	Isaac Leeser. Founder of Maimonidas College
1806	Stand Watie. Cherokee leader; Confederate officer; served at Wilson's Creek and Pea Ridge
1821	Gustave Flaubert. Novelist
1823	Pierre Cochin. Publisher
1830	Joseph Orville Shelty. General of the Confederacy
1843	Bartholomew McCarthy. Scholar
1847	John Murray. Historian
1849	Peter Collier. Publisher
1857	Lillian Nordica. Soprano
1863	Edvard Munch. Artist

1879	Laura Crews. Actress
1880	Zdenka Fossbinder. Soprano
1887	Kurt Atterberg. Composer
1893	Edward G. Robinson. Actor
1897	Lilliam Smith. Author
1912	Henry Armstrong. Boxer
1938	Connie Francis
1941	Dionne Warwick. Singer
1962	Tracy Austin. Winner of U.S. Open

December 13

1521	Pope Sixtus V (Felice Peretti)
1533	Eric XIV. King of Sweden
1662	Francesco Bianchini. Historian
1677	Antoine Touttee. Translater; editor
1735	Elizabeth Dorsey. Pioneer
1738	Basil Brooke. Patriot
1747	Reichen Ammedon of the Colonies
1748	Abiathar Angel of the Revolution
1749	John Breedlove
1753	Daniel Sailor of the Revolutionary forces
1758	Nehemiah Clossev. Patriot
1761	Joseph Allison. Early settler
1805	Johann Lamont. Astronomer
1811	Friedrich Windischman
1818	Mary Todd Lincoln. First Lady
1828	John Savage. Journalist
1832	Alexander Milton Ross. Botanist
1836	Robert Newell. Journalist
1849	Edmund Zalinski. Inventor
1861	Charles Balestier. Writer
1865	Emma Eames. Soprano
1874	Charles Furlong. Explorer
1897	Drew Pearson. Commentator
1913	Archie Moore. Boxer
1929	Christopher Plummer. Actor
1949	Randy Owen. Singer

December 14

1363	Jean Gerson. Orator; writer
1546	Tycho Brake. Astronomer

1586	Georg Calixtus
1640	Deacon John Tenney
1704	Guilio Cordara. Historian
1620	Samuel Badlam of the Colonies
1739	Pierre Samuels Du Pont De Nemours. Instrumental in promoting the Treaty of 1803 by which Louisiana was sold to the United States
1748	Louis Baussett. Author
1775	Philander Chase. Early religious leader
1794	Erastus Corning. Banker; railroader; mayor of Albany
1801	Gen. Joseph Lane of Mexican-American War; senator; governor of Oregon Territory
1826	Frank Boott. Author
1846	James Coffey. Legislator; judge; renowned for "Coffey's Probate Reports"
1860	Lincoln Bush. Inventor
1866	Roger Fry. Director of Metropolitan Museum of Art
1874	Gertrude Elliott. Actress
1874	Josef Shevinne. Pianist
1883	Jane Cowl. Actress
1889	Lefty Tyler of Boston Braves and Chicago Cubs
1895	King George VI of England
1896	James Doolittle. Led raid on Tokyo
1897	Senator Margaret Chase Smith
1911	Spike Jones
1914	Rosalyn Tureck. Pianist
1932	Charlie Rich. Singer
1935	Lewis Arquette. Actor
1936	Charlie Hays. Race car driver
1938	Janette Scott. Actress
1946	Stan Smith. Pro tennis player

December 15

37	Nero. Roman Emperor
1447	Albert the Wise. Duke of Bavaria
1727	Ezra Stiles
1728	Thomas Crandon of the Colonies
1729	Christopher Braucher. Patriot
1733	Samuel Johnston. Governor of North Carolina
1750	Jacob Towne of the Revolution

1754	Paul Henkel Cooper. Evangelist
1778	Christiane Becker. Actress
1786	Edward Coles. Governor of Illinois
1792	Relief Humphrey. Settler
1793	Henry Carey. Publisher
1820	John Caird
1823	Bernard McQuaid. Missionary; president of Seton Hall College; first bishop of Rochester
1832	Alexandre Eiffel of Eiffel Tower
1834	Charles Young. Astronomer
1852	Antoine Beaquerel. Discoverer of radio activity
1878	James Patrick Noonan
1888	Maxwell Anderson. Playwright
1889	William Hougaard. Architect
1892	J. Paul Getty
1894	Adm. Felix Stump. Pacific Fleet Commander
1899	Harold Abraham. Olympic champion
1902	Adm. Bernard Austin. President of Naval War College
1906	Adm. George Whelan Anderson, Jr. Ambassador to Portugal
1922	Lukas Fass. Composer
1926	Rose Maddux. Singer
1933	Tim Conway

December 16

1485	Catherine of Aragon
1758	Abel Center of the Colonies
1768	Charles McGuire. Missionary
1772	Judge William Caball. Governor of Virginia
1775	François Adrien Baceldieu. Composer
1776	Narcisco Duran. Missionary to California
1790	Leopold I. First King of the Belgiums
1824	Thomas Semmes of Confederate Assembly
1839	Alfons Beelesheim. Historian
1847	Charles Antoine
1851	Peter Gibson Thomson. Industrialist
1857	Edward Barnard. Astronomer
1863	George Santoyana. Writer
1866	Anthony Matre
1877	George Atkinson. Actor
1877	Artur Bodanzky. Conductor

1893	Vladimir Golachmann. Conductor of St. Louis Orchestra
1899	Noel Coward
1902	Rafael Alberti. Painter; poet
1903	Hardy Albright. Actor
1917	Arthur Clarke. Writer
1926	John McCracken. Tenor
1938	Liv Ullman
1953	Gary Richards. Actor

December 17

1619	Prince Rupert. Admiral of the Anglo-Dutch War
1739	Topham Beauclerk of the Colonies
1749	Domenico Cimarosa. Composer
1778	Sir Humphrey Davy. Inventor of safety lamp for miners
1796	T.C. Haliburton. Author of *Sam Slick* series
1807	John Whittier. Poet
1808	George Allen. Author
1809	Joseph Nash. Painter
1815	John Bapst. Missionary. First president of Boston College
1834	John Truckenbrod
1840	Hermann Götz. Composer
1848	Frederic Gleason. Musician
1853	Herbert Tree. Actor
1866	Bertha McConnell Bradt of the American Home Economics Association
1874	MacKenzie King. Prime Minister of Canada
1894	Arthur Fiedler. Conductor
1903	Ray Noble. Bandleader
1903	Erskine Caldwell. Author
1908	Willard Libby. Nobel Prize winner
1910	Spade Cooley. Bandleader
1928	Patricia Plunkett. Actress
1929	William Safire. Author
1943	Lorraine Swalve. Widely acclaimed crafts designer
1962	Edward May. Mathematician

December 18

1567	Cornelius Cornelii
1610	Charles Du Cange. Historian

1620	Heinrich Roth. Missionary
1778	Joseph Grimaldi. Singer
1804	Fitzhugh Lane. Composer
1814	Sarah Bolton. Poet
1815	Adm. William Reynolds
1819	Isaac Thomas Hecker
1829	Abraham Rydberg
1835	Lyman Abbott
1852	George Henry White of Congress
1859	Francis Thompson. Poet
1860	Edward MacDowell. Composer
1863	Archduke Franz Ferdinand
1872	Adm. William Stanley. Diplomat
1886	Ty Cobb of baseball
1898	Fletcher Henderson of jazz
1898	Rose Franken. Author
1899	Joseph Daroff. Philanthropist
1907	Christopher Fry. Playwright
1909	Mona Barrie. Actress
1916	Betty Grable
1916	Doug Fraser
1917	Ossie Davis. Actor
1950	Jamie Fricke

December 19

1714	John Winthrop. First American astronomer
1731	Thomas Willing. First president of the Bank of the United States
1749	John Beatty of Revolutionary War
1753	John Gilman. Governor of New Hampshire
1757	Friend Dickinson. Patriot
1790	William Parry. Explorer
1799	Joseph Cretin. First bishop of St. Paul
1817	Jean Danela. Violinist
1845	Charles Adams. UTE Indian agent
1852	Albert Michelson. Nobel Prize winner
1855	Vincent Wehrle. Missionary; first bishop of Bismarck
1858	Horace Troubel. Editor
1865	Minnie Fiake. Actress
1882	Walter Braunfels
1885	William Howard Bishop. Missionary

1888   Fritz Reiner. Conductor
1894   Ford Fricke. Baseball Commissioner
1901   Oliver La Farge. Novelist
1902   Ralph Richardson. Actor
1915   Edith Piaf. Singer
1926   Bobby Layne of football
1933   John Adair. Writer
1934   Al Kaline of Baseball Hall of Fame
1949   Claudia Kolb. Olympic swimmer

December 20

1576   Bl. John Sarkander. Martyr
1692   Martha Potter. Pioneer
1720   Benjamin Loxley of the Colonies
1759   Preserved Buffington. Patriot
1785   James McHenry. Novelist
1798   John Wood. Governor of Illinois
1808   Andrew Johnson. President of the United States
1816   Elijah Dee
1816   William Cooper Nell. Historian
1817   John Laughlin. First bishop of Brooklyn
1824   Calvert Vaux. Architect
1832   Thomas Becker. First bishop of Wilmington
1841   Theodore McMeecham
1865   Elsie De Wolfe. Actress
1868   Harvey Burrill. Botanist
1870   Marie Cahill. Actress
1871   Henry Hadley. Composer
1876   Walter Adams. Astrophysicist
1881   Branch Rickey of Baseball Hall of Fame
1892   Yvonne Armand. Actress
1902   Mary Mildred Pickle. Author
1904   Irene Dunne
1911   Hortense Calisher. Writer
1949   William Jones. Legislator

December 21

1118   St. Thomas a Becket. Archbishop of Canterbury; killed by the King's men

1271	Hawise de Muscgras
1648	Tommaso Ceva. Mathematician
1668	Charles Tournon
1744	Anne Vallayer. Painter
1755	Leonard Bleecher. Patriot
1790	James Grahame. Historian
1799	George Spencer. Missionary
1821	Gustave Flaubert
1862	Jean Vincent. Immunization pioneer
1863	Zelia De Lusson. Operatic soprano
1869	Amos Reusser, M.D. Family doctor of the Year 1948
1874	Gertrude Lane. Editor
1880	John Rainey of Congress
1891	John W. McCormack. Speaker of the House; forty-three year congressman
1892	Rebecca West. Novelist
1892	Walter Hagen of golf
1898	Ira Bowen. Astronomer
1905	Anthony Powell. Novelist
1911	Josh Gibson of Baseball Hall of Fame
1917	Alicia Alonso. Ballerina
1918	Donald Regan. Secretary of Treasury; Chief of Staff
1937	Claire Motte. Ballerina
1940	Frank Zappa. Composer
1944	Michael Thomas. Conductor
1959	Florence Griffith Joyner. Sprint champion

December 22

1095	Roger II. King of Sicily. Founder of the Kingdom of Sicily
1669	John McConibe
1671	Hepzibah Hazen
1696	James Oglethorpe. Founder of Georgia
1723	Karl Abel. Musician
1757	Joshua Severance. Patriot
1770	Demetrius Gallitzen. Missionary
1788	Nathan Wheeler Seeley
1821	Josiah Bushnell Grinnell. Founder of Grinnell University
1829	Sebastian Herbolsheimer
1843	Prentiss Ingraham. Writer
1852	Opie Read. Novelist

1853	Teresa Carreño. Concert pianist
1858	Giacomo Puccini. Composer
1867	William Haroran
1869	Edwin Robinson. Poet
1876	Filippo Marinetti. Playwright
1876	Maynard Smith
1883	Edgard Varese. Composer
1890	Henry Grimmelsman. First bishop of Evansville
1898	Frank Allenty. Actor
1901	André Kastelanetz. Conductor
1905	Kenneth Rexrath. Poet
1908	Giacomo Manzue. Sculptor
1925	Lady Bird Johnson. First Lady
1923	Fernando Corena. Singer
1934	Dave Pearson. Race car driver
1944	Steve Carlton of St. Louis Cardinals, Philadelphia Phillies, San Francisco Giants; Chicago White Sox, Cleveland Indians and Minnesota Twins
1948	Steve Garvey

December 23

1646	Jean Hardwin. Historian
1648	Robert Barclay. Writer
1657	Hannah Dustin. Colonial heroine
1682	James Gibbs. Architect
1708	Ann Elizabeth Weaver. Patriot
1717	John Baegert. Missionary
1722	Axel Cronstedt. Minerologist
1732	Sir Richard Arkwright. Inventor
1734	Thomas Griffin Peachey. Early Virginian
1817	Warren Evans. Writer
1823	Thomas Scott. President of Pennsylvania Railroad
1828	Achille Desurmont. Writer
1830	Charlotte Bernard. Composer
1840	Robert Gifford. Artist
1853	William Henry Moody. Secretary of Navy; Supreme Court Justice
1853	René Bazin. Novelist
1862	Connie Mack of baseball
1869	Sir Hugh Percy Allen. Organist; conductor
1870	Edwin Sabin. Historian

1870	Joseph Maria. Painter
1887	Vincent Bressi. Builder of tunnels, dams, power plants and flood control projects
1889	Adm. Daniel Barbey. Commander of the Seventh Fleet
1900	Otto Soglaw. Cartoonist
1901	Danny Taylor of Washington Senators, Chicago Cubs and Brooklyn Dodgers
1909	Maurice Denham. Actor
1918	José Greco. Dancer
1924	Donald Thorman. Editor
1960	Simon Brookings. Actor

December 24

1167	John Lackland. King of England; signer of Magna Carta
1624	John Houghton of New England
1743	Jacob Reisdorf of the Revolutionary War
1745	William Paterson. Governor of New Jersey
1761	Selim III
1772	Barton Stone
1791	Eugene Scribe. Playwright
1798	Adam Mickiewicz. Poet
1803	Karl Drabusch. Composer
1804	John Stadler
1809	Kit Carson
1822	Matthew Arnold. Poet
1824	Peter Cornelius. Composer
1877	Louis Dumas. Composer
1875	Emile Nelligan. Poet
1881	Juan Jiminez. Poet
1881	Charles Cadman. Composer
1885	Paul Manship. Sculptor
1887	Louis Jouvet. Director
1888	Lucrezia Bori. Soprano
1901	John Coleman. Governor of New York Stock Exchange
1903	Joseph Cornell. Artist
1905	Howard Hughes
1909	Lyle Ball. Painter
1911	Bernice Just. Journalist
1919	Pierre Soulages
1922	Ava Gardner

1931 Jill Bennett. Actress
1941 Howden Ganley. Race car driver

December 25

4BC - 3AD   OUR LORD AND SAVIOR JESUS CHRIST
1400   John de Sutton VI. Constable of Chin Castle; Lord Lieutenant of Ireland
1564   Johannes Buxtorf
1642   Sir Isaac Newton
1717   Pope Pius VI (Gianangelo Broschi)
1737   Bartholomew Dandridge of the Colonies
1761   William Gregor. Discovered titanium
1762   Michael Kelly. Musician
1769   Polly Heath
1781   William Nicholson. Painter
1800   John Phillips. Geologist
1808   Stephen Rowan. Port Admiral of New York; Chairman of Lighthouse Board
1808   Adm. Stephen Clegg of Seminole and Mexican War
1815   Louis Schaefer. Member of Canton City Council, school board and County Commission
1817   Samuel Sloan. President of Delaware, Lackawana & Western Railroad and of Hudson River Railroad. Senator
1819   Eliza Greatorex. Artist
1821   Clara Barton of Red Cross
1829   Patrick Gilmore. Musician
1833   Agnes Robertson. Actress
1838   John Forest. Bishop of San Antonio
1845   Julia Barrett
1850   Isabella Crawford. Poet
1858   Herman Devries. Opera singer
1865   Evangeline Booth
1865   Fay Templeton. Actress
1870   Helene Rubenstein
1873   Otto Hunziker. Scientist
1874   Lina Cavalieri. Opera singer
1883   Maurice Utrillo. Painter
1893   Robert Ripley. Cartoonist
1899   Humphrey Bogart
1899   Ralph Soyer. Artist

1901	Princess Alice. Duchess of Gloucestor
1914	Adm. Noel Gaylor. World War II Fighter Ace
1934	John Ashley. Actor
1934	Giancarlo Baghetti. Racer of Alfa Romeos and Fiats
1948	Barbara Mandrell

December 26

1194	Frederick II. Holy Roman Emperor; crowned in 1220; led a Crusade to the Holy Land in 1228-29
1640	Dr. Thomas Rodman
1716	Thomas Gray. Poet
1723	Friedrich Melchior Grimm. Writer
1738	Thomas Nelson, Jr. Signer of the Declaration of Independence
1751	St. Clement Mary Hofbauer
1754	Ebenezer Huntington. Congressman
1785	John Raffeiner. Missionary
1792	Charles Babbage. Mathematician
1837	George Dewey. American Naval officer; defeated Spanish Fleet at Manilla Bay
1839	Chandler Moulton. Ohio legislator
1844	James Reid. Agricultural expert; developed Reid's Yellow Dent Corn
1853	Wilhelm Dörpfeld. Architect
1854	Eva Toppan. Author
1863	Charles Pathe of Pathe Newsreel
1878	Isaiah Bown
1883	Maurice Utrillo. Artist
1891	Henry Miller. Writer
1894	Jean Toomer. Writer
1907	Elisha Cook, Jr. Actor
1913	Adel Faulkner
1914	Richard Widmark
1924	Glenn Davis. "Mr. Out" of Army
1927	Alan King
1938	Jesse Walters. Chief Justice of Idaho Court of Appeals
1942	Judith Mashburn. Designer

December 27

1350	John I of Aragon
1571	Johannes Kegler. Astronomer

1579	Fr. Andrew White. "Apostle of Maryland"; celebrated first Mass in Maryland
1599	Gabriel Bucelin
1717	Pope Pius VI (Giovanni Braschi)
1737	Ven. Giuseppi Pignatelli
1745	John Choute. Patriot
1748	James Avery of the Colonies
1753	Joshua Beall of Revolutionary War
1773	George Caylay
1794	William Campbell Preston
1795	Ezekiel Piper
1816	Gen. Ellakim Scammon of Civil War
1816	Michael Domenec. Missionary
1822	Louis Pasteur. Founder of microbiology
1828	William Poynts
1835	Adele Douglas (Mrs. Stephen A. Douglas)
1859	Henry Hadow. Writer
1860	Felix Dreyschack. Pianist
1863	Louis Lincoln Emmerson. Governor of Illinois
1875	Lars Boe. Legislator; commander of Order of St. Olav
1886	William Galvin
1896	Louis Broomfield. Novelist
1900	Hans Stuck. Race car driver; winner of German Mountain Championship
1901	Marlene Dietrich
1928	Arlene Lind
1928	John West. Leader of American Research Expedition to Mt. Everest
1930	Wilfred Shhed. Renowned Writer
1935	Larisa Latymina. Gymnast
1941	James McGee of computer fame
1941	John Amas. Actor
1943	Tuesday Weld
1946	Claude Wistrowski. Author
1961	Kevin Edward. Poet

December 28

1584	Juan Dicastillo
1653	Mary Howard. Pioneer
1741	Waterman Clift of the Colonies

1749	Joseph Bangor
1789	Catharine Sedgwick. Novelist
1797	Charles Hodge. Editor of *Princeton Review*
1802	Theodor Ratisbonne
1804	Alexander Johnston. Geographer
1815	Kunigunda Schmidt
1829	John Bannon. Military chaplain of the Confederacy at Corinth and Vicksburg; Confederate commissioner to Ireland
1829	Patrick Gilmore. Musician
1845	Homer Bartlett. Historian
1856	Woodrow Wilson. President of the United States
1865	Felix Yallotton. Painter
1866	George Carter. Governor of Hawaii
1882	Sir Arthur Eddington. Astronomer
1888	F.W. Murnau. Director
1896	Roger Sessions. Composer
1905	Cliff Arquette. Actor
1905	Earl Hines. Jazz pianist
1919	Emily Neville. Novelist
1920	Steve von Buren of Pro Football Hall of Fame
1953	Tovah Fedshue of Broadway
1959	David Mozentale. Actor

December 29

1570	Wilhelm Lamormaini
1639	Anna Gillette. Pioneer
1695	Jean Pater. Artist
1735	Thomas Banks. Sculptor
1756	Anthony Shomo of the Colonies
1766	Charles MacKintosh. Designer of raincoats
1779	Tomasso Bernetti
1800	Charles Goodyear. Inventor of vulcanized rubber
1805	Asa Packer. Legislator; judge; railroader
1807	Nathan Rice. Editor
1808	Andrew Johnson. President of the United States
1809	Ulick Bourke. Scholar; writer
1865	William Adams Brown
1876	Giuseppi De Luca. Museum
1876	Pablo Casals. Violinist
1879	Gen. Billy Mitchell

1920 Viveca Lindfors. Actress
1934 Gerald Felando. Legislator
1937 Mary Tyler Moore
1947 Ted Danson. Actor
1953 Gelsey Kirland. Ballerina
1957 Carol Perkins
1959 Martin Moran. Actor
1965 Amanda Green of Broadway

December 30

41 Titus. Roman Emperor
1642 Vincenzo Filicaia. Poet
1748 Adam Bahn of the Colonies
1754 Asa Atwood. Patriot
1754 Samuel Cutts. Served in Revolutionary War
1784 Stephen Long. Explorer
1830 Francis Drake. President of Indiana, Illinois & Iowa Railroad; governor of Iowa
1847 John Peter Altgeld. Governor of Illinois
1847 John Marshall Hamilton
1855 Katharine Bush. Author
1865 Rudyard Kipling. Poet
1884 Karl David. Composer
1900 Clarence Barnhart. Lexicographer
1906 Carol Reed. Director
1910 Paul Bowles. Composer
1910 Adm. Norvelle Word. Vietnam War leader
1920 Michael Allinson. Actor
1934 Richard Aslanian. Conductor
1934 Fred Lorenzen. Race car driver
1935 Sandy Koufax
1938 Susan Maxman. Architect
1942 Bo Nance of football
1944 Peter von Reichbauer. Statesman
1975 Tiger Woods

December 31

1491 Jacques Cartier. Explorer
1514 Andress Visalius. Anatomist

1550	Henri I de Lorraine
1720	Charles Stuart. "Bonnie Prince Charlie"
1743	Nathan Longfellow of the Colonies
1738	Gen. Charles Cornwallis. Surrendered to Gen. Washington at Yorktown
1755	Nicholos Holma. Mathematician
1783	Thomas Macdonough. Navy commander in War of 1812. Victor at Battle of Lake Champlain
1811	Thaddeus Amat. Bishop of Monterey-Los Angeles
1815	George Meade of the Battle of Gettysburg
1817	James Field. Publisher
1820	Mary Ann Sadlin. Author
1838	Aime Jules Dalou. Sculptor
1842	Giovanni Baldini. Painter
1844	Ebe Tunnel. Governor of Delaware
1846	Cassius McGregor
1855	Giovanni Pascali. Poet
1858	Harry New. Postmaster General
1859	Jeff Abugon. Writer
1862	Robert Cuddihy. Publisher of *Literary Digest*
1864	Robert Aitken. Astronomer
1869	Henri Matisso. Painter
1885	Norreys O'Conor. Poet
1898	Frank Skinner. Composer
1904	Nathan Mistein. Concert violinist
1920	Rex Allen. Actor
1928	Ross Barbour. Singer
1943	John Denver
1946	Diane Von Furstenberg. Designer
1968	Joelle Marie. World renowned singer; has appeared in Soviet Union, Poland and at major American events

Bibliography

Academic American Encyclopedia
Almanac of Famous People
American Ancestry
American Leaders
Ancestral Roots of Sixty Colonists

Burke's Presidential Families of the United States of America

Catholic Encyclopedia
Celebrety Who's Who
Current Biography
Cyclopedia of American Biography

DAR Patriot Index
Dictionary of American Catholic Biography
Dictionary of American Religious Biography
Directory of Notable Americans

Encyclopedia Americana
Encyclopedia of American Business History and Biography
English Origens of New England Families

Famous American Admirals

Genealogies of Long Island Families
Genealogy of New England
Great Lives from History
Great Writers of the English Language

History of Broome County
History of Bureau County, Illinois

International Dictionary of Film and Filmmakers

Liberty's Women

Motion Picture Almanac

National Encyclopedia of American Biography, The
New Century Cyclopedia of Names
New Encyclopedia Brittanica
Notable American Women

Readers Encyclopedia of American Literature
Register of Maryland's Heraldic Families

Theatre World

Unabridged Blue Water Bridge History, The

Who's Who in America
Who's Who in American Politics
Who's Who in Racing
Who's Who in the West
Who's Who in the World
Who Was Who in America
Who Was Who in American Politics